Modern Vacuum Practice

Nigel S. Harris MSc, CPhys, MInstP

Manager, Vacuum Training Centre
Edwards High Vacuum International

The McGraw-Hill Companies

London · New York · St Louis · San Francisco · Auckland
Bogotá · Caracas · Lisbon · Madrid · Mexico · Milan
Montreal · New Delhi · Panama · Paris · San Juan · São Paulo
Singapore · Sydney · Tokyo · Toronto

Published by
McGRAW-HILL Publishing Company
Shoppenhangers Road, Maidenhead, Berkshire, SL6 2QL, England
Telephone Maidenhead 01628 23432; Fax 01628 770224

British Library Cataloguing in Publication Data
Harris, Nigel S.
    Modern vacuum practice
    1. Vacuum technology
    I. Title
    621.5′5
    ISBN 0-07-707099-2

Library of Congress Cataloguing-in-Publication Data
Harris, Nigel S.
    Modern vacuum practice/Nigel S. Harris
      p. cm.
    Bibliography: p.
    Includes index.
    ISBN 0-07-707099-2
    1. Vacuum technology. I. Title.
    TJ940.H37   1989
    621.5′5—dc19

McGraw-Hill

A Division of The McGraw·Hill Companies

Printed and bound in Great Britain at the University Press, Cambridge

Printed on permanent paper in compliance with ISO Standard 9706

To
Sheila, Tina and Mark

Contents

Preface

Modern vacuum practice

My original intention when I started this work was to produce a revised second edition of my 80-page booklet *Vacuum Engineering*, which first appeared in 1977. I had intended to keep the framework of the booklet unchanged, but to make additions, corrections and updating.

However, on closer inspection it was apparent that there were some subject areas within the old text that had seen important recent developments—particularly developments in vacuum pump design to combat the demanding physical problems met, for example, in the semiconductor manufacturing industry. Missing from the text was a thorough explanation of how gases in a vacuum system could be identified using mass spectrometry. Missing from vacuum measurement were the newer gauges such as the capacitance manometer types and the spinning rotor gauge. Missing also was any reference to gauge calibration; the different types of valves, safety, etc., and very little information was given about the newer vacuum pumps such as cryopumps and turbomolecular pumps.

The task was therefore to be a more major one, to bring together the wide-ranging aspects of *modern* vacuum technology, but at what level? A review of the books already published showed that in the sixties a large number of textbooks were published dealing with vacuum technology: many are now out of print; those remaining are out of date. Very few new books dealing with the subject have appeared since then; those that have are either specialist (dealing with topics such as ultra-high vacuum or process vacuum system design) or have perhaps treated the subject at undergraduate and graduate level.

It became clear that there was a need for a simple, non-mathematical, technician-level, up-to-date book—one that helps the reader understand the most widely used methods of obtaining and measuring low pressure and that includes some of the major recent developments in production and measurement. Additionally, such a book needed to include hints and tips on system design and on the care and operation of system components, in order to be more practical and give the reader more confidence in working with vacuum equipment. Some of this information is available from vacuum equipment manufacturers' sales literature, working instructions and review articles, and technical publications such as *Vacuum* are a useful source of information. However, such information, while available from a variety of sources, is not apparently available in one simple-level reference work. Thus the seed of an idea for a book dealing with *modern vacuum practice* was born.

Who would be interested in such a book? My experience with my earlier work, together with my full-time involvement since 1976 in training in vacuum technology and vacuum equipment maintenance, led me to believe that a wide audience of technicians, engineers and scientists would find the book useful. The aim would be to provide a good background/reference text for technologists, technical assistants and development engineers in the industries and government establishments that now require staff with vacuum experience. These industries include those involved in semiconductor manufacture, freeze drying, vacuum metallurgy and a host of coating and research applications.

The approach has been to assume little or no prior vacuum knowledge. The treatment, except for Chapter 13, is non-mathematical, with the arithmetic used extending only to the application of decimals, powers of ten and logarithms (all of which are explained in the Appendices). Vacuum terminology is in most cases explained as it appears, and follows that found in Parts 1 to 3 of ISO document 3529, *Vacuum Vocabulary* (1981).

Essentially the book concentrates on methods of producing and measuring pressures into the high vacuum range and on how systems are designed, constructed and operated. Emphasis on rotary pump and diffusion pump systems is deliberate since such systems are those most commonly encountered. Some of the logic behind the operation and troubleshooting of such systems can be extended to other types of pumping systems. Leak detection is of importance to the worker who has just constructed a vacuum system. Chapter 14 considers the different methods available but concentrates on the different uses of the mass spectrometer type of leak detector since this instrument has become widely accepted in industry to inspect a wide variety of products. Safety has been included in a guide to the safe use of vacuum equipment (Chapter 15), a very important topic in our health- and safety-conscious society. It is, I believe, the first time that a chapter has been devoted to this subject as part of a vacuum textbook.

Where reference is made to routine maintenance of some of the various system components, it is imperative in all cases to first consult the manufacturer's instructions before carrying out any service work and to adhere strictly to their recommended procedures.

With regard to personal safety, in some sections brief reference is made to suggested ways of cleaning/looking after vacuum equipment. Normal safe working procedures should be observed when using cleaning chemicals. No smoking should be permitted in cleaning areas. Safety glasses should be worn at all times and, in addition, when appropriate, a plastic face guard, plastic apron and rubber gloves should be worn.

Acknowledgements

I would like to express my gratitude to the Directors of Edwards High Vacuum International for their encouragement to write this book and for their permission to reproduce many of the illustrations that have been used during training courses held in the Edwards Vacuum Training Centre.

The author is indebted to his colleagues Dr Steve D. Hoath and Dr Neil Tenwick, who in their own valuable time were good enough to read the first draft of the manuscript and have made many useful suggestions and contributions since. Thanks are also due to those who have contributed to the content without specific reference or acknowledgement, and those who have helped in the preparation of the manuscript.

N. S. Harris

1

Introduction

1.1 What is a vacuum?

The vacuum chamber shown in Figure 1.1 is a leak-tight vessel, to which is connected a pump capable of removing the air from within the chamber. As air is removed various levels of vacuum can be obtained depending upon how much air is withdrawn.

There is less air contained in a chamber under vacuum than when that chamber is open to the atmosphere.

However, a vacuum cannot be defined as 'a space completely devoid of any gaseous material' since, even with the best vacuums ever created, it is not possible to remove all the gas. The word gas is often used loosely in vacuum practice to denote both permanent gases such as hydrogen, oxygen and nitrogen and the inert gases such as neon and argon as well as vapours.

The distinction between a gas and a vapour is a subtle one and is considered in more detail in Chapter 2.

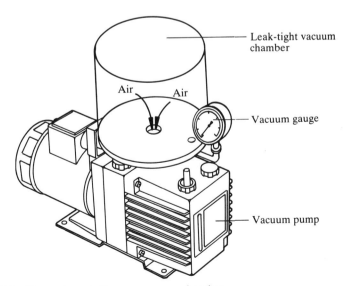

Figure 1.1   Removing air from a vacuum chamber

1

A gas is made up of many small invisible particles called molecules, which are moving about rapidly, with speeds typically about 1600 km h^{-1} (just over 1000 mile h^{-1}), in all directions. At atmospheric pressure the number of molecules contained in a cube of air having 1 cm sides is about 25 million million million. At the lowest pressure ever obtained artifically in a chamber this number is reduced to about 350 molecules in the same volume. In the lowest pressure regions in intergalactic space this number is estimated to be about 4 per cubic metre.

Gas molecules, unlike those of the liquid and solid phases, are not 'bound' to one another, but are freely moving about, filling all space available to them. If we visualize these molecules by imagining them as tennis balls moving about rapidly in all directions in a room, they collide frequently with each other and with the walls, floor, ceiling, etc. This striking and rebounding results in a 'push' or force on all the surfaces within the room. The average force on a unit area, say one square metre, is called the pressure.

$$\text{Pressure on a surface} = \frac{\text{Force}}{\text{Area}}$$

In a similar way atmospheric air applies a pressure at the earth's surface by the continuous collision of gas molecules. Although the atmosphere is without walls, it is confined from below by the earth's land–sea surface and from above by the force of gravity that prevents its outward escape.

In the case of atmospheric air at sea level,

Pressure = 14.7 pound-force per square inch (psi or lbf in^{-2})

= 1 tonf ft^{-2} (approximately)

= 1 kgf cm^{-2} (approximately), etc.

If the crushing force due to the atmosphere is so large, why are we normally unaware of it? The answer is that the pressure inside our bodies is almost the same as that outside. A similar balance exists with other objects, unless air is removed from within them; then the effect of atmospheric pressure may be noticeable. An example of this occurs when using a drinking straw to drink liquid from a bottle. When you suck, your lungs expand and air passes into them from the straw. Atmospheric pressure pushing down on the surface of the liquid in the bottle is now greater than the pressure of the air in the straw and so forces the liquid up to your mouth.

If a vacuum chamber has a pressure of 1 kg on every square centimetre of its outer surface when it has a vacuum inside it, its constructional material will therefore have to be able to withstand these sorts of pressures; otherwise it may collapse. Typically, thick-gauge stainless steel or mild steel are used, or in some cases glass. The vacuum inside the chamber can be described as an environment where the pressure is less than that of the immediate surrounding atmosphere.

A vacuum is a space in which the pressure is below atmospheric pressure.

In vacuum technology, we are concerned with the pressure inside the chamber. With large amounts of gas inside, the pressure is high and the vacuum is poor (rough); with most of the gas removed, the pressure is low and the vacuum is improved, of 'high' quality. Thus the reference pressure for the best vacuum is zero pressure, and all other vacuum conditions will be positive pressures.

However, in some applications, especially of rough vacuum, some workers may be concerned with forces produced by the pressure difference between atmosphere and the chamber. This leads to the use of 'gauge' pressures, which are zero when this difference is zero but can go negative when the chamber is evacuated (depressurized) and become positive when pressurized. However, in vacuum technology we are concerned with absolute pressure within the chamber, and vacuum gauges display this pressure and not the pressure differential across the chamber walls.

Air consists of a mixture of gases. Over 99 per cent of dry atmospheric air is nitrogen and oxygen. All other gases make up less than 1 per cent. Table 1.1

Table 1.1   Composition of dry air

Gas	Percentage by volume
Nitrogen	78.08
Oxygen	20.95
Argon	0.93
Carbon dioxide	0.03
Neon	0.001 8
Helium	0.000 5
Methane	0.000 2
Krypton	0.000 1
Hydrogen	0.000 05
Xenon	0.000 008 7

shows the composition of dry air as a percentage by volume. Note that since the major gas constituent is nitrogen, many gauges used for measuring the pressure in a vacuum system are calibrated for dry nitrogen and are accurate under normal vacuum conditions where small quantities of other gases and water vapour are present. When other gases predominate, perhaps in a system where a process gas is being introduced into the chamber, corrections to the readings may be needed.

1.2 Units of vacuum measurement

The degree of vacuum is not normally expressed as the number of molecules present in a known volume; nor is it normal to use psi as a unit of measurement.

Vacuum measurement units such as the ones listed below have been used in the past. However, they are now no longer acceptable as units of pressure within the European Community, although some of these units, particularly the torr, are still favoured in some parts of the world (especially in the USA).

millimetres of mercury	(mmHg)
micrometres (microns) of mercury	(μmHg)
inches of water	(inH$_2$O)
torr or millitorr (mtorr)	

Such units relate to early pressure measurements made with manometers of the liquid-column type. The liquids were either mercury or water. (Note that Hg is the chemical symbol for mercury as H$_2$O is the chemical symbol for water and that 'm' is often used as an abbreviation for milli or one-thousandth of a whole part; similarly 'μ' is an abbreviation for micro or one-millionth of a whole part.)

The derivation of the mmHg units resulted from experimental work done by an Italian called Torricelli in 1644. He filled a long tube (closed at one end) with mercury and inverted it into a bowl of mercury (see Figure 1.2). Torricelli found that the mercury flowed out of the tube until the weight of the column was balanced by the pressure exerted on the surface of the mercury by the air above. In other words, the weight of the mercury in the column equalled the weight of a similar-diameter column of air that extended from the ground to the top of the atmosphere. The height of the mercury column was about 760 mm high. Torricelli noted that when air pressure increased, the mercury in the tube rose; conversely, when air pressure decreased, so did the height of the column of mercury. The length of the column of mercury, therefore, became the measure of air pressure—the column was in fact the first barometer. Standard atmospheric pressure at sea level equals 760 mmHg (29.92 inHg).

A Frenchman, Pascal, reasoned that atmospheric pressure should decrease with height. Unfortunately he suffered from poor health and could do no mountain climbing himself, so he sent his brother-in-law up the Puy-de-Dôme with a Torricelli barometer in 1646. In climbing just over $1\frac{1}{2}$ km (1 mile), the mercury column dropped about 76 mm (3 inches).

In climbing to the top of Mount Everest one would experience a reduction in the atmospheric pressure conditions to about one-fourth of that existing at sea level. The fact that the air is rarified at high altitudes was borne out in a

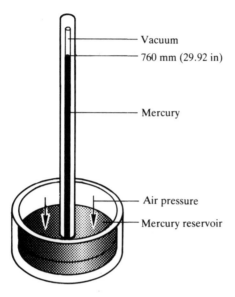

Torricelli barometer

Figure 1.2   The weight of a column of mercury is balanced by the pressure exerted on the surface of the mercury in the reservoir

dramatic way by some nineteenth century balloonists attempting to study the upper atmosphere. Several were known to have passed out upon reaching heights in excess of 6 kilometres and some even perished in the rarified upper air. As one progresses further out into space this blanket of air becomes thinner and the pressure encountered becomes less. The rate at which pressure decreases with altitude is not a constant, i.e. the rate of decrease is much greater near the earth's surface. The pressure is reduced by approximately one-half for each 5 km increase. Hence, at 5 kilometres the pressure is one-half its sea-level value, at 10 kilometres it is one-fourth, at 15 kilometres it is one-eighth, and so on.

At 16 km (10 miles) the pressure is around 77 mmHg and 90 per cent of the atmosphere has been traversed. At 100 km (62 miles) only 0.000 03 per cent of sea-level pressure remains, the pressure here being about 0.000 02 mmHg. At an altitude of around 160 km (100 miles) the pressure is about 0.000 001 mmHg while at about 480 km (300 miles) this has fallen to 0.000 000 01 mmHg. Around 1000 km (620 miles) the pressure is 0.000 000 000 1 mmHg. Traces of our atmosphere thus extend for thousands of kilometres beyond the earth's surface. To say where the atmosphere ends and outer space begins is very arbitrary.

As far as the production of artificial vacuum is concerned, until the seventeenth century there were no effective vacuum pumps. Then in 1654

Otto von Guericke, who was Mayor of a town called Magdeburg in Germany, carried out a series of experiments using a pump that he had invented. This pump was a piston pump using wet leather seals (similar to a bicycle pump). In one experiment he used his pump to remove the air from two large hollow metal hemispheres, fitted together to give an airtight sphere. So good was his pump that two teams of eight horses, attached to each half of the sphere, could not separate the sphere. Pumps like Guericke's were capable of attaining pressures of about 10 mmHg and with modifications were in use until the end of the nineteenth century (see the example shown in Figure 1.3). The emerging electric lamp industry at the end of the nineteenth century stimulated the invention of other types. These used mercury as a liquid piston, and although slow and cumbersome in use they produced more perfect vacua than the earlier types, achieving pressures much lower than 0.001 mmHg. In 1905, a new period was ushered in by the invention by Gaede of the rotary mercury vacuum pump, which on account of its simplicity and speed of working was an immediate success. With the arrival of these pumps, a smaller unit of pressure was used, and the 'micron of Hg' came into use. A micron is a unit of length, one-thousandth of a millimetre.

The term 'mmHg' was later replaced by 'torr' in honour of Torricelli. Thus

$$1 \text{ mmHg} = 1 \text{ Torr}$$

$$1 \text{ micron Hg} = 1/1000 \text{ or } 0.001 \text{ mmHg} = 0.001 \text{ Torr} = 1 \text{ mTorr}$$

Standard atmospheric pressure = 760 mmHg = 760 Torr

The standard metric unit of pressure under the Système International d'Unités (SI) is the newton per square metre ($N \text{ m}^{-2}$), which also takes the

Figure 1.3 A pair of Magdeburg hemispheres and a vacuum pump used at Leipzig in 1709 to demonstrate the effects of a vacuum. (By kind permission of Museum of Mathematics and Physics, Dresden, Germany)

name pascal. In this notation a standard atmosphere has a value of 101 325 pascals (the pascal is equal to 0.007 5 Torr).

As from January 1978, the torr ceased to be an acceptable unit of pressure in Europe (although it is still in use in the USA). The alternatives were the pascal and the millibar (mbar). Although the latter is not related to the SI unit by a preferred factor of 10^3, the majority of the European vacuum industry has adopted the mbar as the most acceptable unit. Millibars are used almost exclusively in meteorology and the lines of equal pressure (isobars) seen on most weather maps are in these units (see Figure 1.4). The base unit, the bar, is equal to 100 000 N m^{-2} or pascals (Pa). These latter units are too small to be used conveniently in many pressure applications. Note that

$$1 \text{ bar} = 1000 \text{ mbar} = 750 \text{ Torr}$$

$$
\begin{aligned}
\text{Standard atmospheric pressure} &= 101\,325 \text{ pascal (N m}^{-2}) \\
&= 1.013\,25 \text{ bar} \\
&= 1013.25 \text{ mbar} \\
&= 101.325 \text{ kPa}
\end{aligned}
$$

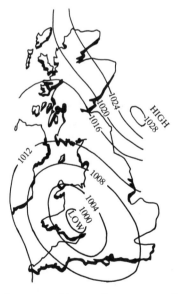

Figure 1.4   Isobars as shown on this weather map are lines connecting places of equal barometric pressure. The units of pressure measurement used here are mbars (millibars)

Table 1.2 gives a series of conversion factors for the various vacuum units discussed. For an explanation of numbers written as powers of ten see Appendix C.

Table 1.2 Conversion factors for vacuum pressure units

	mbar	torr (mmHg)	micron (μmHg)	pascal (N m^{-2})	Atmospheres
mbar	1	0.75	750	100	9.87×10^{-4}
torr (mmHg)	1.33	1	10^3	133	1.32×10^{-3}
micron (μmHg)	1.3×10^{-3}	10^{-3}	1	0.133	1.32×10^{-6}
pascal (N m^{-2})	10^{-2}	7.5×10^{-3}	7.5	1	9.87×10^{-6}

To convert a unit in the left-hand column to a unit in the top line multiply by the relevant factor.

1.3 The vacuum spectrum

The practical vacuum spectrum here on earth extends from 1013 mbar down to the lowest pressures so far produced artificially, which are claimed to be of the order of 10^{-14} mbar (as we shall see it is difficult to measure such low pressures). However, such values are not as low as some found in nature. It is thought that pressures of 10^{-16} mbar exist in interstellar space and of around 10^{-22} mbar in intergalactic space. Figure 1.5 shows how pressure changes with altitude over the pressure range that we are going to consider. It is convenient to separate this wide vacuum spectrum into ranges:*

Low or rough vacuum:	1013 mbar–a few mbar
Medium vacuum:	a few mbar–10^{-3} mbar
High vacuum:	10^{-3} mbar–10^{-7} mbar
Ultra-high vacuum (UHV):	below 10^{-7} mbar

Using this terminology we can say, for example, 'this pump will produce a low pressure' or 'this pump will produce a high vacuum'; both are correct.

1.4 Production of vacuum

The perfect vacuum system consists of a completely leak-tight vacuum chamber attached to a perfect pump which operates at constant volume flow rate (constant 'pumping speed') down to the lowest pressure required. Any molecule passing through the pump inlet port would be captured and not reemitted; the pump would not contribute any contaminating gases or

* From ISO 3529: *Vacuum Technology—Vocabulary*, Part 1: *General Terms* (1981).

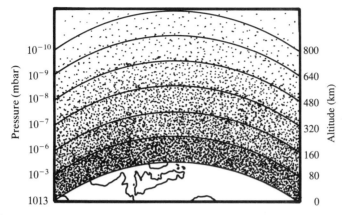

Figure 1.5   How pressure varies with altitude. (1 km = 0.62 mile)

vapours to the system atmosphere. It would also not influence the system being pumped, with stray electric fields, magnetic fields, energetic particles or any similar influences. It would require simple services and not be sensitive to their failure and would also be easy to maintain, compact in size, cheap to purchase and run, and simple to operate. No existing single type of pump has all these ideal characteristics, although some do approach the ideal in some aspects of their performance.

For example, real pumps differ from the ideal pump in relation to constant 'speed' and pumping effectiveness for different gases. Effective pump 'speeds' are obtained only over a restricted pressure range; they may be relatively constant over some of this range.

The principle types of vacuum pumps can be classified into two groups:

1. *Gas transfer pumps.* These pumps remove gas molecules from the pumped volume and convey them to a higher pressure in one or more stages of compression. Examples are the rotary-vane pump, rotary-piston pump, mechanical booster (Roots pump), diffusion pump, turbomolecular pump.
2. *Capture pumps.* In these, gases are not passed directly to another type of pump but are retained (permanently or in the short term) in the pump itself by sorption or condensation etc. on internal surfaces. Examples are the sorption pump, sublimation pump, sputter-ion pump, cryopump.

Figure 1.6 shows the normal operating pressure range of the pumps to be covered. It can be seen, for example, that for applications involving rough- and medium-vacuum equipment, several types of pumps are available, including sorption pumps and various combinations of mechanical rotary pumps. Depending on design, several types of mechanical rotary pumps exist, e.g. rotary-vane, rotary-piston and mechanical booster. Mechanical boosters (to be dealt with in Chapter 6) are normally used in combination with either

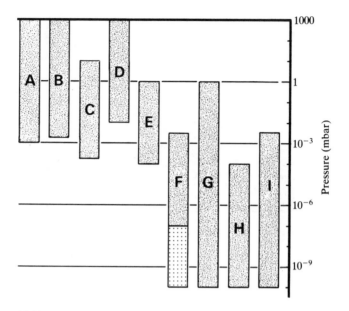

KEY	PUMP
A	Sorption
B	Mechanical rotary
C	Mechanical booster (normally used with a rotary pump)
D	'Dry pump'
E	Vapour booster
F	Diffusion (low pressures obtained with accessories)
G	Turbomolecular
H	Ion
I	Cryo

Figure 1.6 Typical vacuum pump operating ranges between atmospheric pressure and 10^{-10} mbar

the rotary-vane or rotary-piston pump to increase their effective speed and extend the working range. Diffusion and turbomolecular pumps are examples of high vacuum pumps which must be 'backed' continuously by a mechanical pump to exhaust the gases to the atmosphere.

Some vacuum pump types

Low and medium vacuum
Pumps such as the rotary-vane, rotary-piston and mechanical booster are all positive displacement types. They reduce the pressure in a system by repeatedly taking samples of gas into the pump. The pump mechanism isolates the gas from the inlet, compresses it and then expels it through an outlet.

High and ultra-high vacuum

Diffusion pump　　Here gas transport is achieved by a series of high-velocity vapour jets (normally oil vapour is used) emerging from an assembly within the pump body. In normal operation a portion of any gas arriving at the inlet jet is entrained, compressed and transferred to the next stage.

Turbomolecular pump　　This contains a rotor with inclined blades moving at high speed between corresponding stationary blades in a stator. Gas molecules entering the inlet port acquire a velocity and preferred direction superimposed on their thermal velocity by repeated collisions with the fast-moving rotor. Rotational speeds for small pumps are typically $60\,000$ rev min^{-1}.

Cryogenic pump　　Operation is achieved by the condensation, freezing and/or sorption of gas at surfaces maintained at extremely low temperatures, thus removing them from the gas phase in the vacuum system.

Sputter-ion pump　　This makes use of the gettering principle, in which a cathode material (usually titanium) is vaporized—or sputtered by bombardment with high-velocity ions. The active gases are pumped by chemical combination with the sputtered titanium, the inert gases by ionization and burial in the cathode, and the light gases by diffusion into the cathode.

1.5 Vacuum applications

The first major use of vacuum technology in industry occurred about 1900 in the manufacture of electric light bulbs. Other devices requiring a vacuum for their operation followed, such as various types of radio valves. It was discovered that certain processes carried out in a vacuum achieved either superior results or ends actually unattainable under normal atmospheric conditions.

Such developments included the 'blooming' of lens surfaces to increase light transmission, the preparation of blood plasma for blood banks and the production of reactive metals such as titanium. The advent of nuclear studies in the fifties provided impetus for development of vacuum equipment on a large scale. Increasingly applications for vacuum processes were steadily discovered. More recently, the growth of the microelectronics industry and the various space research programmes have given a new quantitative jump to vacuum technology, due to the numerous vacuum problems that had to be solved.

Reasons for using a vacuum

A process or physical measurement is generally performed in a vacuum for one or more of the following reasons:

1. To remove active atmospheric constituents that could cause a physical or chemical reaction (e.g. in electric lamps, during melting and sintering of metals, packaging of dairy products and in encapsulation of electronic components).
2. To achieve a pressure difference (e.g. for holding, lifting and transportation, and in vacuum forming of plastics).
3. To decrease energy transfer (e.g. for thermal and electrical insulation).
4. To remove occluded or dissolved gas or volatile liquid from the bulk of material (e.g. degassing of oils and freeze drying).
5. To extend the distance that a particle must travel before it collides with another, thereby helping the particles in a process to move without collision between source and target (e.g. vacuum coating; in cathode ray, X-ray and television tubes; ion implantation; particle accelerators; and in electron beam welding).
6. To produce clean surfaces (useful in the preparation of pure, thin films and in surface studies such as corrosion, catalysis and tribology).

Some of the above applications use vacuum technology as a step in production, the final product being used in atmospheric conditions (vacuum coating, freeze drying, etc.). Others result in products in which the vacuum exists during the useful life of the product (lamps, TV tubes, etc.).

Applications in industry

Because of the wide range of industrial processes that vacuum technology covers and the even wider range with which it is connected only a few examples can be dealt with here.

In the electrical engineering industry, power transformers and cables are seriously impaired by the presence of water. During manufacture of transformers, large quantities of water need to be removed from the core and windings, and this is usually achieved by a combination of heating and application of vacuum. Vacuum also removes air trapped in the windings, which are finally impregnated with oil. The removal of the air ensures efficient impregnation.

Improvements in electrical insulation can be found in the manufacture of certain types of high-voltage or high-current-carrying switches in which arcing or flashover are reduced in a low-pressure environment.

In the chemical engineering and allied industries, vacuum processing is usually required to lower the boiling point to enable compounds to be separated into their individual chemical components. In chemical works vacuum may be used in purging pipelines. Vacuum may also be used for the concentration of slurries, pastes and concentrates by water removal. In the metallurgical industry, manufacturers of special steels and alloys and producers of fine powders for sintering etc. may use a vacuum melting, casting, annealing, heat treatment and pouring technique. Aircraft engine manufacturers and repairers and aircraft component manufacturers use vacuum brazing techniques.

Freeze drying
A large application is in freeze drying of heat-sensitive pharmaceutical and biological preparations, finely divided reactive powders, museum specimens and tissue for microscopy. Freeze drying is the dehydration by sublimation from the frozen state. It has the advantages of minimal product spoilage due to elimination of liquid and because of processing at sub-zero temperatures. Products treated using this method include coffee, fruit juices, vegetables and meat, blood plasma and antibiotics, such as penicillin. Even bone and arteries have been preserved for long periods.

Vacuum coating
Vacuum coating is used in many manufacturing fields. Under high vacuum, metals and salts can be vaporized and the vapours condensed on any solid surface. In the simplest case, the evaporation of aluminium, a small length of aluminium wire is looped onto a tungsten filament. When the filament is heated in a vacuum to bright red heat, the aluminium melts and freely evaporates. Work to be aluminized is arranged to face the aluminium source positioned some distance away; it is then coated with a thin film of the metal. In this way it is possible to produce mirrors, ophthalmic lenses, antiglare and antistatic glasses, television tubes and decorative plastics. The method is used for continuous roll coating of plastic sheet and paper.

An additional new application in vacuum deposition emerged during the seventies with the increased used of architectural glass. In order to improve the heat insulation properties in buildings in cold countries or to protect against intense sun radiation and contribute to energy saving in hot countires where buildings are air-conditioned, coating of glass has become a large-scale production technique. Large-area glass coating is now carried out under vacuum conditions with multiple layers deposited, which are both transparent and heat reflecting.

Vacuum leak testing
In the aerospace, electronics, atomic energy, cryogenics and refrigeration industries, vacuum leak testing of components is of great importance. For the detection of potential gas or liquid leaks into or out of components, the vacuum mass spectrometer leak detector is found to be the most sensitive and easiest to use instrument. Using helium as a search gas the instrument can

find, for example, holes of approximately one-millionth of a centimetre in diameter in a component having a wall thickness of 2 mm.

In refrigeration system manufacture, there are two reasons why the system should be leak-tight. The system consists of a closed circuit of tubing around which a volatile fluid is circulated by means of a compressor. If sufficient refrigerant fluid escapes, the system will cease to function, and if air leaks into the system, atmospheric moisture will collect and freeze in the cold zone of the circuit, causing a blockage. Other examples of the detector's uses can be found on oil rigs, testing large heat-exchangers and pipes; finding leaks in large road tankers carrying liquefied gases; checking sealed devices such as transistors, crystals and relays; leak detection of equipment associated with uranium-enrichment projects in the nuclear industries and in the armaments industry for missile construction.

Semiconductor industry

The semiconductor industry came into being when it was discovered that the properties of large and complex electrical circuitry could be duplicated in minute form by depositing layers in amazingly accurate configurations on tiny silicon chips. These chips are produced in large quantities from circular wafers which, having been processed, are cut up into the small autonomous chips. The chips are subsequently included in circuit boards to form complete electrical circuits.

The technology for producing semiconductor devices has now become firmly established and production is proceeding world-wide on a large and rapidly increasing scale. From the initial stage of silicon crystal growing to producing specific semiconductor properties, through the multiple fabrication stages involving deposition and etching, a controlled vacuum environment is required. Most of the processes have in common the need to admit one or more process gases into the vacuum chamber for the generation of a plasma and the promotion of any chemical reactions required. Chemical reactions may produce gaseous or solid reaction products.

Pumping systems for semiconductor processes may therefore have to accommodate process gases, gaseous reaction products and particulates. Depending on the process, the gaseous materials may be explosive, flammable, aggressive, corrosive or toxic while solids are likely to be abrasive. These products present difficult operational problems which have to be overcome to ensure that acceptable reliability is achieved.

Successful pumping systems employ vacuum pumps incorporating features that assist in combating aggressive materials together with accessories to trap these materials or mitigate their effect. Choice of pumping fluid and methods of operation are also important.

The following lists illustrate a few examples of applications which have been classified in terms of the degree of vacuum required:

1. *In industry*

Rough-vacuum range near atmospheric to about 1 mbar:

Mechanical handling

Vacuum packing and forming

Gas sampling

Filtration

Figure 1.7 Vacuum packaging of meat in a food factory. (Note the mechanical booster pump under the workbench.) (By kind permission of Farmers Boy Limited, Bradford)

Figure 1.8 A pumping 'cart' for exhausting and sealing colour television tubes under vacuum. (The cart consists of a rotary pump and diffusion pump system.) (By kind permission of Philips Components, Durham)

Degassing of oils
Impregnation of electrical components
Semiconductor device fabrication

At lower pressures down to about 10^{-4} mbar:
Refrigeration dehydration
Metallurgical processes: i.e. melting, casting, sintering, brazing
Chemical processes: vacuum distillation and freeze drying
Semiconductor device fabrication

Pressures down to about 10^{-6} mbar:
Cryogenic and electrical insulation
Lamps, television tubes, X-ray tubes, etc.
Decorative, optical and electrical thin-film coating
Mass spectrometer leak detectors

Figure 1.9 An ion-implanter which can be used to modify the electrical characteristics of semiconductors. (Note the cryopump and gate valve.) (By kind permission of Whickham Ion Beam Systems Limited, Newcastle upon Tyne)

2. *In research*

Pressure down to about 10^{-6} mbar:
Electron microscopes
Analytical mass spectrometers
Particle accelerators
Large-space simulation equipment

Pressure region down to and below 10^{-9} mbar:
Thermonuclear experiments
Field ion and field emission microscopes
Storage rings for particle accelerators
Clean surface studies
Specialized space simulator experiments

The photographs in Figures 1.7 to 1.9 show some of the applications mentioned.

Some relevant physical concepts

In order to understand vacuum phenomena, an appreciation of some basic physics is necessary. The following section sets out some of the basic concepts which will be referred to in later chapters.

2.1 States of matter

Matter consists of very large numbers of small, discrete particles known as atoms. Atoms may combine together in chemical reactions to form arrangements termed molecules. For example, two hydrogen atoms may combine with one oxygen atom to make a molecule of water. Certain atoms—those of the so-called 'inert' gases such as helium, neon and argon—are very unlikely to combine chemically, and the atoms remain single and unattached. The atoms of other gases such as hydrogen, nitrogen and oxygen rarely exist as single atoms, and form diatomic gas molecules, i.e. molecules of two identical atoms. For simplicity it can be considered that all matter can be divided into three states, or phases, these being the solid state, the liquid state and the gaseous state. Some matter exists in each of these states; e.g. ice, water and steam are three states of the same substance. There are solid, liquid and gaseous forms of carbon dioxide etc.

The solid state

In a solid, the atoms are bound tightly together in fixed positions relative to each other by interatomic forces. Solids therefore have fixed volumes. Heat energy contained in the solid appears as vibrations of the atoms about these positions. Normally these vibrations are not strong enough to break the bonds holding the atoms in place. If heat is supplied to the solid, the vibrations increase in strength and eventually the bonds break, causing the solid to melt and form a liquid. A few substances will convert directly from the solid to the vapour without passing through the intermediate liquid stage. The effect is known as sublimation and is the basis of vacuum freeze drying.

The liquid state

In a liquid, the atoms (or molecules) have no fixed positions, and are in constant random motion, wandering freely about the bulk of the liquid.

Because the atoms are still very close together, they are still influenced by interatomic binding forces, particularly at the surface. If heat is added to the liquid, the random motions of the atoms become more vigorous, eventually reaching the point at which they move quickly enough to overcome the binding forces. The liquid then boils and becomes a gas.

The gas phase

In a gas, the molecules are on average very much further apart than in solids or liquids. For example, in air at atmospheric pressure and room temperature, about 0.01 per cent of the space is actually occupied by the molecules. In a solid such as copper, the figure is 74 per cent. In a gas, the molecules are constantly moving in random directions, but because they are further apart interatomic forces have very little effect. At very short intervals, individual molecules collide with and bounce off others (about 10 000 000 000 times per second for each molecule at atmospheric pressure). A gas thus automatically expands to fill a volume into which it is led.

2.2 Vapours and saturated vapour pressure

Some molecules have sufficient energy to escape from the surface of bulk solids and liquids and become part of the surrounding atmosphere. Raising the temperature of the bulk substance will increase the rate of escape. A dish of water, for example, if left uncovered in the open air, will evaporate; vapour molecules rapidly diffuse away from the parent liquid and in general produce what is known as an unsaturated vapour. On the other hand, if the liquid is in an enclosed vessel there is a high probability that a free molecule will collide with the liquid surface and be recaptured (i.e. condensation). Thus, an equilibrium will be established between evaporation and condensation and the air is unable to accept any more water; it has become *saturated* and the pressure it exerts is called the saturated vapour pressure. Since the rate of evaporation falls with a decrease in temperature, the saturated vapour pressure also decreases. Refrigerated surfaces are frequently built into vacuum systems to cause vapours to condense on them and act as pumps. This is known as 'cold trapping' or 'cryopumping'.

A saturated vapour is one in equilibrium with its liquid.

In vacuum, the rate of evaporation increases. Molecules leaving the surface have less chance of colliding with air molecules, and evaporation proceeds more rapidly. If the pressure is progressively reduced from atmospheric, the rate of evaporation gradually increases until, at a certain value of pressure, evaporation becomes much more rapid—for our example of a dish of water, the water begins to boil at room temperature (20 °C). This occurs at a pressure of 23.4 mbar.

As long as any water exists at room temperature anywhere in a vacuum system, the minimum pressure attainable in the system will be 23.4 mbar (the saturated vapour pressure). A little water can produce very large volumes of vapour, and hence can take a long time to be pumped away. Note that rapid evaporation tends to result in a lowering of the temperature of the water (unless sufficient heat is available), and it may freeze. At 0 °C the saturated vapour pressure of ice is 6.11 mbar.

The same process applies to any liquid inside the vacuum system. The implication of this is that great care must be taken to avoid contaminating a system (e.g. by careless handling (fingerprints) or unclean work (fluxes, greases, cleaning fluids, etc.)) with possible vapour sources.

All materials, including solids, have vapour pressures, although some are very low at normal temperatures and others may be a problem for a vacuum system. Table 2.1 gives vapour pressures of various liquids at 20 °C.

Table 2.1 Vapour pressure of various liquids at 20 °C

Liquids	Vapour pressure (mbar) at 20 °C
Organic liquids	
Acetone	250
Benzene	100
Carbon tetrachloride	120
Ethyl alcohol	60
Isopropyl alcohol	510
Toluene	30
Trichlorethylene	80
Xylene	7
Metals	
Mercury	1.5×10^{-3}

Materials having a noticeable vapour pressure at the highest intended operating temperature must obviously be avoided. The vapour pressure of most metals is so low that it does not restrict their use for vacuum applications. However, alloys that contain zinc, lead and cadmium, for example, have unsuitably high vapour pressures for vacuum application. Cadmium is commonly used to plate steel screws and brass has a zinc content. Table 2.2 shows the vapour pressure of 'difficult' metals at various temperatures. Vapour evolved from the hot metals is likely to condense on adjacent cooler surfaces. A potential problem can be a metal film condensed on an electrical insulator, thus causing a short circuit.

Table 2.2 Saturated vapour pressure versus temperature of certain metals

	10^{-5} mbar	10^{-4} mbar	10^{-3} mbar
Aluminium	710 °C	800 °C	880 °C
Cadmium	145 °C	175 °C	215 °C
Zinc	205 °C	245 °C	285 °C

2.3 Boyle's law

If a given mass of permanent gas is confined in a cylinder by an airtight piston and the volume of the gas is decreased by the movement of the piston, the pressure of the gas increases. When the volume of the gas increases the pressure decreases. Gases are therefore compressible. Robert Boyle investigated the relationship between gas pressure (p) and volume (V) and deduced his law.

> Boyle's law states that for a fixed mass of gas at a given temperature, the product of the pressure and volume is constant; i.e. pV is constant.

Figure 2.1 illustrates the effects of Boyle's law. A graduated cylinder is fitted with a piston, pressure gauge and thermometer. In Figure 2.1a, a

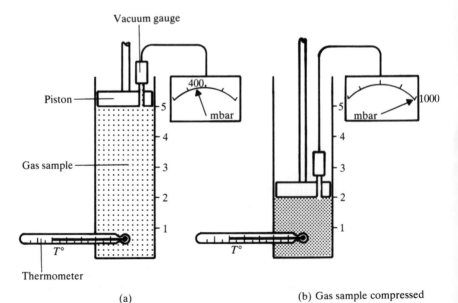

(a)

(b) Gas sample compressed

Figure 2.1 An illustration of Boyle's law

sample of gas occupies a volume of 5 litres, the pressure is 400 mbar and the temperature is $T°$. In Figure 2.1b, the gas has been compressed to 2 litres. The quantity of gas remains the same, i.e. none has escaped or leaked in. If we assume that the temperature remains constant, Boyle's law can be used to predict the new value of pressure. The original pressure times original volume is equal to the final pressure times final volume:

$$P_{final} = \frac{400 \times 5}{2}$$

$$= 1000 \text{ mbar as indicated on the gauge}$$

Note that the McLeod gauge utilizes gas compression and Boyle's law to obtain a pressure reading.

If an unsaturated vapour is compressed at constant temperature, its pressure and volume change approximately in accordance with Boyle's law until the saturated vapour pressure is reached. Further attempts at compression cause vapour to condense. Thus, under conditions of saturation, Boyle's law does not hold. This behaviour, however, does not occur at all temperatures. If the compression process is carried out at successively higher temperatures, eventually a temperature is reached (known as the critical temperature) at and above which the vapour will not condense, no matter what pressure is applied. The so-called permanent gases (e.g. H_2, O_2, N_2 and inert gases) have critical temperatures well below normal room temperature. On the other hand, the vapours of water, mercury and the common organic cleaning liquids have critical temperatures above normal room temperature and hence can be liquefied by compression.

One of the implications of this is that certain pumps and the gauge mentioned above depend upon high gas compression for their action, and any associated condensation of liquids may produce a deterioration in the pump's performance or errors in gauge reading. These effects will be discussed in the relevant sections.

2.4 Processes occurring at a boundary wall

If all the gas to be removed from a vacuum chamber were located in the volume of the chamber, it could be removed easily by a pump in a very short time. However, in practice this will never happen. The lowest pressure obtainable and the time taken to pump-down a vacuum system are limited by the slow evolution of additional gases and vapours from the inside surfaces of the vacuum chamber and from the surfaces of any components within it. This surface gas release is the result of several processes including vaporization, desorption, diffusion and permeation (see Figure 2.2). In the absence of leaks, these processes, together with possible gas emission from the pumps, determine the composition of the residual atmosphere in the system.

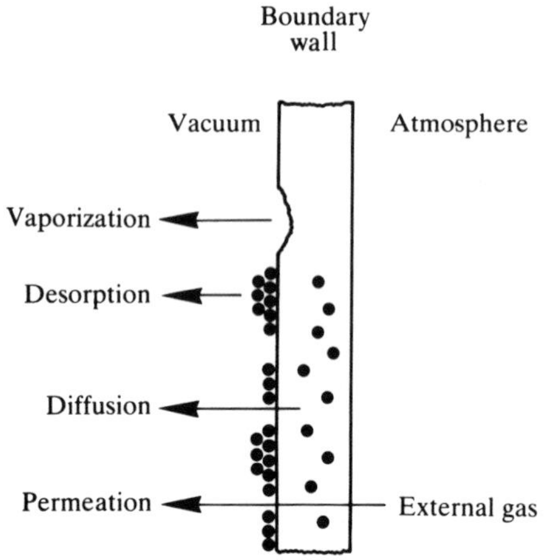

Boundary
wall

Vacuum ⎢⎢ Atmosphere

Vaporization ◄────────

Desorption ◄──────

Diffusion ◄──────────

Permeation ◄──────────── External gas

Figure 2.2   Gas and vapour load associated with the boundary wall

Desorption

Desorption is the liberation of gases and vapours taken up (sorbed) by a material. *Outgassing* is the spontaneous desorption of gas from a material (see Section 13.5). *Degassing* is the deliberate desorption of gas from a material, i.e. an accelerated outgassing process, e.g. by using heat.

The reverse process, the taking up of gas or vapour by a solid or liquid is known as sorption. This can occur as *adsorption,* in which the gas or the vapour is retained at the surface of the solid or the liquid, or by *absorption,* in which the gas diffuses into the bulk of the solid or liquid. Where the sorption is due to physical forces, in which no definite chemical bonding occurs, it is known as physisorption. Sorption in which the formation of chemical bonding occurs in the process is known as chemisorption.

Adsorbed gases and vapours are not bound for ever, but there is a possibility that they are released back into the gas space and are replaced by others. The time they stay on the surface is called residence time. The release of adsorbed gases is usually very rapid and with rising temperature the mean residence time is further reduced. Therefore a temperature rise accelerates the desorption of adsorbed gases.

Absorbed gases must first diffuse from the substance to the surface, from where they are discharged to the gas space. Here, again, the desorption rapidly rises with rising temperature. In the high vacuum and ultra-high vacuum range, the pumping times are mainly determined by desorption, as well as other factors such as permeation.

Diffusion

Diffusion is a process in which particles (atoms or molecules) move through a solid, liquid or gas. Examples of gas diffusing from the wall of a vacuum system container include hydrogen from aluminium and from iron and also carbon monoxide and carbon dioxide from mild steel. Diffusion is a much slower process than desorption; the rate of transport through the bulk to the surface governs the rate of release into the vacuum.

Permeation

In addition to desorption, the permeation of gases through the wall and through sealing materials can have a certain influence in high and ultra-high vacuum. An example is helium permeation through glass or through elastomer seals. Permeation involves several steps, simplified as follows:

1. Impact of the gas atoms or molecules on the outer surface
2. Adsorption
3. Movement of the gas in the wall material from the saturated outer surface layer due to a concentration gradient
4. Transfer of the gas to the inner surface of the wall
5. Desorption of the gas from the inner surface

Permeation is a function of the temperature, type of gas and structure of the material.

2.5 Gas mixtures—partial pressures

The gas in a vacuum chamber usually consists of a mixture of various constituent molecules. Simply, each gas constituent can be considered to exert a pressure appropriate to the number of its molecules present. The total pressure of the mixture will be equal to the sum of the partial pressures of the mixture:

$$P_{total} = P_1 + P_2 + P_3 + \cdots \text{ etc.}$$

Partial pressures as low as 10^{-15} mbar have been measured using instruments known as mass spectrometers. There are several forms of the instrument but the general principle of operation is common to them all; the gas molecules are ionized, accelerated, separated into groups according to their masses and finally collected. Their operation and use will be dealt with in Chapter 4.

2.6 Ionization

Gas molecules consist of groups of atoms bound together by relatively strong forces so that the molecular entity stays intact throughout all the inter-molecular collisions. The atom as a whole is electrically neutral and gases are therefore extremely poor conductors of electricity. They can be made better conductors by altering the balance of the positive and negative charges in the

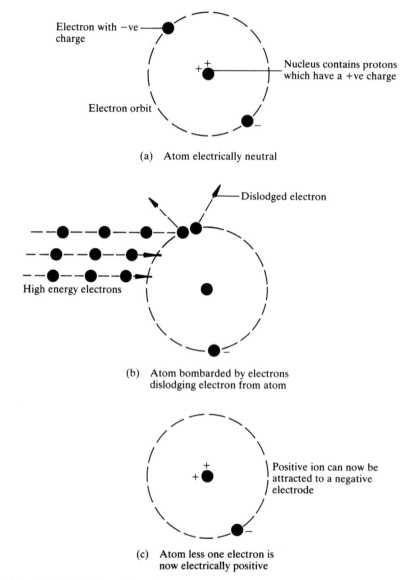

Electron with −ve charge

Nucleus contains protons which have a +ve charge

Electron orbit

(a) Atom electrically neutral

Dislodged electron

High energy electrons

(b) Atom bombarded by electrons dislodging electron from atom

Positive ion can now be attracted to a negative electrode

(c) Atom less one electron is now electrically positive

Figure 2.3 Ionization of an atom

atoms. Atoms can be considered to be composed of three parts: namely electrons, protons and neutrons. The electrons are negatively charged particles moving in orbital paths around the nucleus of the atom (see Figure 2.3a). The nucleus contains two particles: electrically neutral neutrons and positively charged protons whose charge is equal but opposite to the charge on the electrons.

An example of ionization occurs in the mass spectrometer. Gas enters a low-pressure region where it is bombarded with electrons from a heated filament. The bombardment causes electrons to be stripped from the atoms. By removing one or more electrons from their orbits, the atom can take on an overall positive charge. Such a charged atom is termed a positive 'ion' (see Figure 2.3b and c). Once a gas has been ionized the motion of the ions can be influenced by electric and magnetic fields. Note that an atom that gains an electron forms a negative ion, but this is less common.

The symbol for a positive ion is written as, say, N^+, in this case a nitrogen atom with one electron removed, or O^{2+}, an oxygen atom with two electrons removed, i.e. doubly ionized.

A space containing ionized gas will conduct electricity by means of the transport of ions between electrodes (see Figure 2.4). Positive ions will be attracted by the negative voltage on the right-hand electrode (cathode) and move towards it. When an ion reaches the electrode it takes up an electron from it. The ion and electron combine to make a neutral atom. The current of ions through the gas is thus accompanied by a flow of electricity in the circuit which will show as a deflection on a meter. The electric current relates directly to the number of ions in the space and therefore to the number of molecules. This simple arrangement (called a discharge tube) forms the basis of ionization vacuum gauges.

An ion is an atom or group of atoms which have become charged, positively or negatively, through losing or gaining electrons.

The ability to move gas atoms by applying chosen electric (and magnetic) fields to ions finds several applications in vacuum technology.

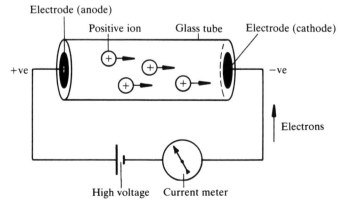

Figure 2.4 Discharge tube. Positive ions move towards the cathode.

2.7 Mean free path, molecular density and monolayer formation time

In Section 1.3, we saw how the vacuum spectrum was split into different ranges. Although the limits of the regions look arbitrary they do tend to correspond to a different physical situation. In order to describe these situations it is useful to utilize concepts such as mean free path, molecular density and monolayer formation time.

Mean free path

The molecules in a gas are in constant random motion, periodically colliding with one another and moving off in new directions. The distance travelled by a molecule between one collision and the next has an important bearing on various vacuum phenomena.

The average distance travelled by a molecule between collisions is termed the mean free path. The length of the mean free path depends on the temperature and pressure of the gas and the size of the molecules. For air at room temperature, a simple formula can be used to calculate it:

$$\text{Mean free path} = \frac{6.4 \times 10^{-3}}{P} \text{ cm}$$

where P = pressure in mbar

In this case, then, the mean free path is about 6×10^{-6} cm at atmospheric pressure and 64 metres at 10^{-6} mbar.

Molecular density

Molecular density was described briefly in Section 1.1 and is the average number of molecules per unit volume.

Time to form a monolayer

This is the time taken for a clean surface in a vacuum to be covered by a layer of gas of one molecule thickness.

Figure 2.5 shows how these physical characteristics vary with pressure. In the rough and medium vacuum ranges, the number of gas phase molecules is large compared with those covering the surface. In the high vacuum range the gas molecules in the vacuum system are located principally on surfaces, and the mean free path equals or is greater than the relevant dimensions of the enclosure. In the ultra-high vacuum range, the time to form a monolayer is equal or longer than the usual time for laboratory measurements; thus 'clean' surfaces can be prepared and their properties can be determined before the adsorbed layer is formed.

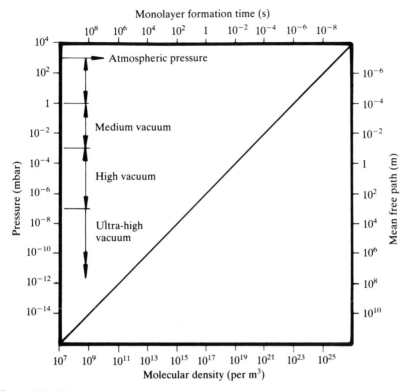

Figure 2.5 Values of molecular density, mean free path and time to form a monolayer, as a function of pressure, for air at 25 °C

2.8 Volume flow rate (pump speed)

The rate at which gas flows across a plane, whether through a leak, along a pipeline or into a pump, is generally described in terms of volumetric flow rate. The volume flow rate of gas transported across a plane is the volume of gas (at a specified temperature and pressure) crossing that plane in a given interval of time, divided by that time. The unit of flow rate is normally expressed in cubic metres per second ($m^3 s^{-1}$).

When applied to the performance of vacuum pumps this term is commonly referred to as the 'pump speed' or 'pumping speed'. Pumping speed (S) is normally measured in litres per second ($l s^{-1}$), cubic feet per minute ($ft^3 min^{-1}$) or cubic metres per hour ($m^3 h^{-1}$). Conversions can be made between these units using Table 2.3.

The value of pumping speed quoted by the pump manufacturer is normally measured at the pump inlet and is determined by methods laid down by the International Organization for Standardization (ISO), Pneurop (a coordinated

Table 2.3 Conversion of volume flow rate units

	$l\,s^{-1}$	$ft^3\,min^{-1}$	$m^3\,h^{-1}$
$l\,s^{-1}$	1	2.12	3.60
$ft^3\,min^{-1}$	0.472	1	1.70
$m^3\,h^{-1}$	0.278	0.589	1

To convert a unit in the left-hand column to a unit in the top line multiply by the relevant factor.

assembly of manufacturers of compressors, vacuum pumps and pneumatic tools from 12 European countries) or the American Vacuum Society (AVS).

2.9 Throughput (Q)

The volume flow rate (pump speed) gives no information about the actual quantity of gas which is flowing, since the density of the gas—which is highly variable—does not enter into the measurement. Clearly a 1 litre per second flow of gas at 10^{-6} mbar contains far fewer molecules than the same flow at 1000 mbar. However, by stating the flow in terms of pressure times volume flow rate units, or 'throughput', the variation of gas density with pressure is allowed for. This gives a flow unit that relates directly to the actual quantity of gaseous matter in the flow (mass flow). The throughput is given by the equation:

$$\text{Throughput} = \frac{\text{pressure} \times \text{volume}}{\text{time}}$$

or

$$Q = \frac{PV}{t} = PS$$

The throughput varies with temperature and is generally specified at room temperature, approximately 20 °C. The unit in common use is the millibar-litre per second (mbar l s^{-1}). Many other units have been used, for which conversion factors are listed in Table 2.4.

Table 2.4 Conversion of throughput units

	$mbar\,l\,s^{-1}$	$Torr\,l\,s^{-1}$	$atm\,cm^3\,s^{-1}$	$lusec$	$atm\,ft^3\,min^{-1}$
$mbar\,l\,s^{-1}$	1	0.75	0.987	7.5×10^2	2.097×10^{-3}
$Torr\,l\,s^{-1}$	1.333	1	1.316	10^3	2.795×10^{-3}
$atm\,cm^3\,s^{-1}$	1.013	0.76	1	7.6×10^2	2.12×10^{-3}
$lusec$	1.333×10^{-3}	0.001	1.32×10^{-3}	1	2.79×10^{-6}
$atm\,ft^3\,min^{-1}$	4.78×10^2	3.58×10^2	4.72×10^2	3.58×10^5	1

To convert a unit in the left-hand column to a unit in the top line multiply by the relative factor.

3

Vacuum measurement

The pressures obtainable by vacuum techniques, and which normally require measurement, span a very wide range from 1013 mbar down to 10^{-10} mbar and below. Pressure measurement is difficult in that there is no single physical effect that can be used over the whole vacuum range. Thus a series of vacuum gauges are available, each of which has a characteristic measuring range. These gauges fall into several main groups: mechanical phenomena gauges, transport phenomena gauges and ionization phenomena gauges.*

The group of mechanical phenomena gauges depend on the actual force exerted by the gas, e.g. U-tube, capsule dial, strain, capacitance manometer and McLeod gauge, etc. Measurements can be made, for example, by measuring the displacement of an elastic material or by measuring the force required to compensate its displacement.

In the main they are absolute gauges in that the determination of pressure does not depend on the gas species. Their use below about 10^{-4} mbar becomes difficult due to the difficulty in accurately measuring minute changes.

Transport phenomena gauges either measure the gaseous drag on a moving body, i.e. the spinning rotor gauge, or depend on thermal conductivity of the gas, e.g. Pirani and thermocouple gauges.

Ionization phenomena gauges ionize the gas and measure the total ion current, e.g. cold cathode ionization (Penning) gauges and hot cathode ionization (ion) gauges.

Both transport and ionization phenomena gauges are *indirect* methods of pressure measurement, where a particular physical property of the gas is measured.

Figure 3.1 shows the pressure spectrum and typical vacuum gauge coverage by gauges readily available today. Some variation in the limits of measurement of these gauges can be found. For example, one manufacturer used to produce a special low-pressure McLeod gauge capable of reading down to 10^{-6} mbar. Most of these gauges measure what is termed *total pressure*, i.e. no distinction is made between the permanent gases (hydrogen, nitrogen, etc.) and vapours (oil vapour, water vapour, etc.). However, under

* From ISO 3529: *Vacuum Technology—Vocabulary*, Part 3: *Vacuum Gauges* (1981).

31

KEY	GAUGE	KEY	GAUGE
A	U-tube manometer	F	Spinning rotor
B	Capsule dial	G	Pirani
C	Strain	H	Thermocouple
D	McLeod	I	Penning
E	Capacitance manometer	J	Ionization

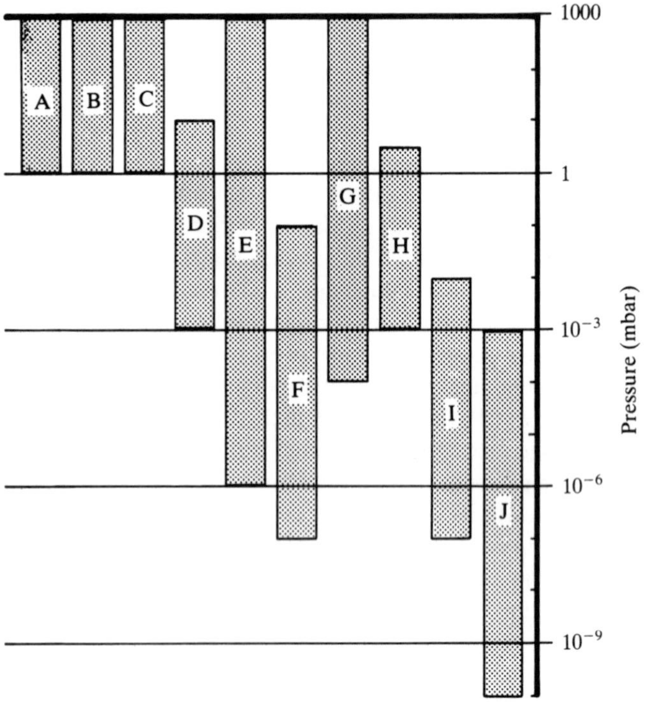

Figure 3.1 Pressure spectrum and typical vacuum gauge coverage

certain conditions the McLeod gauge can give a reading for the partial pressure of non-condensable permanent gases in a sample that contains both gases and vapours. Vapours contained in the gases may be partially condensed, depending upon the compression ratio and the degree to which the vapours are saturated.

There is an obvious difficulty in measuring the true pressure of a gas mixture of unknown composition. This information is best obtained using an instrument capable of determining the partial pressures of separate atomic and molecular species in the vacuum, i.e. a mass spectrometer. Partial pressure measurement will be discussed in detail in Chapter 4.

Figure 3.2 gives a relative cost comparison for the gauges to be considered. The comparison is based on the cost of a capsule dial gauge having 1 unit cost. The spinning rotor and capacitance manometer gauges are shown to be

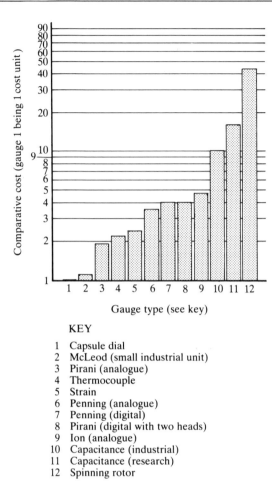

KEY

1 Capsule dial
2 McLeod (small industrial unit)
3 Pirani (analogue)
4 Thermocouple
5 Strain
6 Penning (analogue)
7 Penning (digital)
8 Pirani (digital with two heads)
9 Ion (analogue)
10 Capacitance (industrial)
11 Capacitance (research)
12 Spinning rotor

Figure 3.2 Price comparison of gauges with associated gauge heads taking the cost of a capsule dial gauge as unity

very expensive gauges in comparison with the capsule dial. However, as we shall see, both are precision measuring instruments capable of great accuracy.

To most engineers in industry the most important factor in the assessment of any measurement method is repeatability. For example, if a sensor repeats a calibration curve produced in comparison with a primary standard time and time again to within acceptable limits, then the device is good as a transfer standard. Both the spinning rotor and the capacitance manometer meet these requirements, and so too, to a lesser extent, does the ionization gauge.

The need for accuracy is discussed at the end of the chapter. Let us see how the gauges mentioned work.

3.1 Mechanical phenomena gauges

U-tube manometer

Range: atmospheric pressure to 1 mbar

Figure 3.3a shows a manometer consisting of a U-shaped glass tube containing a liquid, e.g. mercury. One end of the tube is connected to the vacuum system; the other is left open to the atmosphere. When the system is evacuated, the difference in pressure across the U-tube applies a force to the mercury. This force lifts the mercury until the pressure difference is balanced by the difference in the height of the mercury in each leg of the tube.

System pressure = atmospheric pressure − pressure due to height of liquid (h)

The height difference (h) is read off an adjacent scale and this can be converted to millibar pressure units and substituted in the above formula. Note that the reading of the gauge is dependent on atmospheric pressure, which varies from day to day. A modified version of the U-tube having a sealed end is shown in Figure 3.3b. Here the reference pressure is not atmospheric. In manufacture, the tube is first evacuated to less than 10^{-2} mbar and is then filled with mercury; because the end is sealed the system pressure is compared with a constant pressure rather than the variable atmospheric pressure, so permitting greater accuracy. An increase in accuracy can be obtained by inclining the U-tube, thereby increasing the scale length.

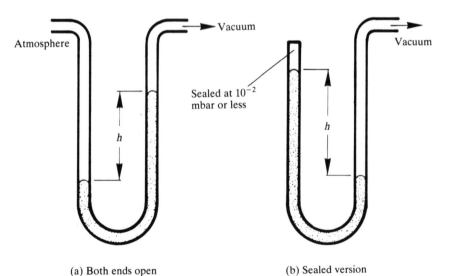

(a) Both ends open   (b) Sealed version

Figure 3.3   U-tube manometer

Capsule dial gauges

Range: 1000 to 1 mbar

The capsule dial gauge consists of a sealed capsule mounted inside a leak-tight case (see Figure 3.4) which can be connected to the system. Movement of the capsule with change of pressure is transmitted via an appropriate mechanism and converted into a movement of a needle traversing a calibrated scale.

The gauge is barometrically compensated; atmospheric pressure plays no part in the reading obtained, the pressure used for the result being the difference between the pressure to be measured and the sealed-off pressure of the capsule. The accuracy is typically ± 5 per cent of full-scale deflection. Typical ranges are 0 to 25, 0 to 50, 0 to 125 and 0 to 1000 mbar. Mechanical gauges of this type are only suitable down to 1 mbar. Below this level, the pressure energy available to produce movement is very small and would require an extremely delicate mechanism to harness it.

If a large quantity of dirt or process material passes through the vacuum system, the gauge, being part of the system, will become contaminated and the accuracy may be impaired. Dirt on the dial is a good indication of contamination. Oil inside the interior may originate from the rotary pump. Maintenance should be limited to cleaning and sealing ring renewal.

Rarely, gauge pointers can become displaced. This may be a result of an excessively rapid rise in pressure caused by the system vacuum being broken too quickly. Normally carefull removal and resetting of the needle is possible, following the maker's instructions.

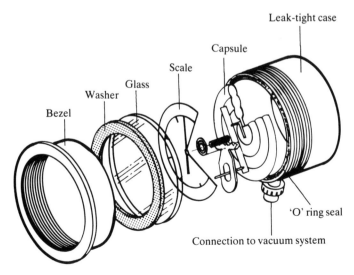

Figure 3.4 Exploded view of capsule dial gauge

In situations where this gauge is mounted a long distance from the point of measurement, there is likely to be an inaccuracy due to the pressure drop along the line. The strain gauge allows remote readings to be made accurately.

Strain gauges

Range: 1000 to 1 mbar

In this gauge a flexible diaphragm, usually made of stainless steel, forms a leak-tight separation between the vacuum system and the surrounding atmosphere (see Figure 3.5). Pressure changes within the gauge head produce a force that deflects the diaphragm. The deflection produces a change in electrical resistance of a semiconductor strain gauge attached to it. The output voltage may be set at atmospheric pressure and at high vacuum to give high accuracy. The response time of this type of gauge is very fast and remote readings up to several metres away can be made.

Capacitance manometers

Range: atmospheric pressure to 10^{-6} mbar

'Capacitance manometer' is a term used to describe a pressure-measuring instrument that electronically senses the deflection of a diaphragm by means of a change in capacitance between an electrode (or electrodes) and the diaphragm as the diaphragm deflects under forces due to the pressure differential across it.

Gauge heads are available with chambers on either side of the diaphragm having connections for differential pressure determinations or with one side permanently sealed under vacuum, at a pressure of around 10^{-6} to 10^{-7} mbar, for absolute pressure measurement.

Figure 3.6a shows an absolute type of double-sided construction containing two fixed electrodes located on either side of the diaphragm. Of the two cavities so formed one is maintained at a reference pressure and the other is at

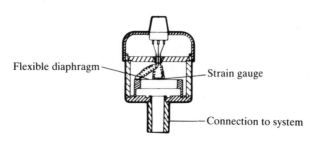

Figure 3.5 Section through a strain gauge head

the pressure to be determined. This method offers excellent zero stability and a high signal-to-noise ratio and is most suited to clean vacuum environments. It is not really suited to applications where the measurement cavity may be subject to particulate and corrosive attack, such as is found in some semiconductor applications. Filters at the entrance to the cavity keep particulate matter from entering the sensor but considerably lengthen the

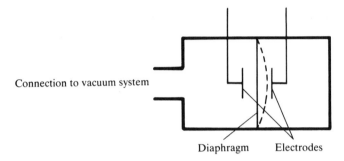

Connection to vacuum system

Diaphragm   Electrodes

(a) Double-sided symmetrical configuration

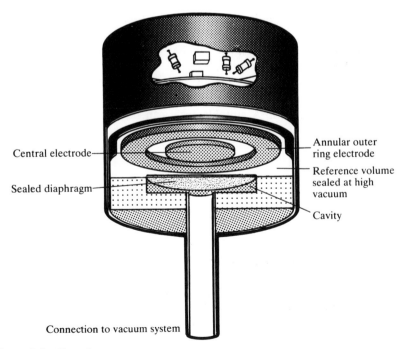

Central electrode

Sealed diaphragm

Annular outer ring electrode

Reference volume sealed at high vacuum

Cavity

Connection to vacuum system

Figure 3.6 Capacitance manometer

response time of the gauge head. Decontamination of the gauge head is difficult because cleaning solvents are difficult to remove.

To avoid these problems a single-sided construction is available (see Figure 3.6b). Here there is no electrode in the cavity volume containing the gas at the unknown pressure. Both electrodes are on the reference side. One fixed electrode is opposite the centre of the diaphragm and is surrounded by an annular reference electrode. The other electrode is the moveable sealed diaphragm.

Capacitance manometers give accuracy typically an order of magnitude better than other gauges, such as Pirani, thermocouple and capsule dial gauges, and are used as transfer standards. They have a fast response. There are few disadvantages in the use of the capacitance manometer as a transfer standard. The only major precaution to be taken, as with any pressure/vacuum calibration, is to minimize the change of temperature while carrying out measurements. For this reason some models of capacitance manometers are available where the head is maintained at a constant 45 °C by a heater.

McLeod gauge

Range: 10 to 10^{-3} mbar

The McLeod gauge utilizes Boyle's law (see Section 2.3) to determine gas pressure within a system. The gauge consists of a system of glass tubes and a glass bulb which are interconnected so that a large volume of gas of the pressure to be measured is compressed into a smaller volume of higher pressure by means of a mercury column. The difference in levels of mercury between the compressed volume and that at system pressure is a function of the pressure in the system. A suitable scale can be mounted showing the pressure in the system as a function of the difference in column height. As the gauge depends only on the known initial volume trapped, the final compressed volume and the pressure in this final volume—all of which can be directly measured—it is an absolute gauge.

A version of the gauge is shown in Figure 3.7. It consists of a pumping line (A) which is connected to the vacuum system, a closed capillary (B) and a parallel capillary (C), a bulb (D) and a scale.

To take a pressure reading the gauge is slowly rotated anticlockwise through 90° from its ready position about the pivot. As this is done the mercury runs towards the capillaries. When it reaches point X, a sample of gas at the prevailing pressure is trapped inside the gauge. The mercury is further raised until it reaches a reference mark near the top of capillary C. This reference mark coincides with the top of the closed capillary B. As the mercury flows into the bulb D and capillary B, the trapped gas is compressed and, if the gas behaves as an ideal gas, will behave according to Boyle's law.

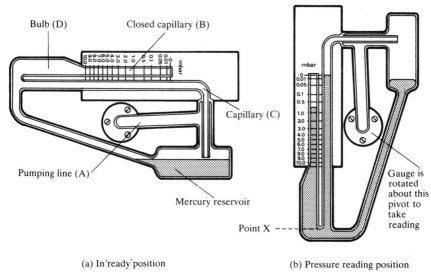

(a) In 'ready' position   (b) Pressure reading position

Figure 3.7 McLeod gauge (vacustat) before and after compression of trapped gas

Due to the low pressure, differences which would not be detectable in a simple U-tube are amplified to the extent that a measurable height difference occurs between the mercury levels in the two capillaries. The pressure is read off the scale mounted alongside the closed capillary. After the reading has been completed the gauge should be returned to its normal position with the capillaries horizontal.

If the sample of gas contains a proportion of a condensable vapour, the reading of the gauge can differ greatly from the true pressure. As the gas sample is compressed, a pressure can be reached at which the vapour condenses into liquid. The volume of the liquid becomes very small compared with that of the vapour, with the result that the partial pressure of the vapour in the original sample is excluded from the measurement. The gauge will then give a measurement only of the partial pressure of the uncondensed gases in the sample.

An illustrative example of where this can occur is when the low-pressure limit (or 'ultimate pressure') of an oil-sealed mechanical pump is being measured. Oil vapour migrates from the pump into the gauge and mixes with any residual gases which the pump is unable to remove. The McLeod gauge is operated and a reading of perhaps 1×10^{-3} mbar is obtained. During operation of the gauge, the oil vapour is condensed and its partial pressure is not measured. The pressure indicated is the partial pressure of the residual permanent gases. If another type of gauge is used, a type which does not condense vapours, a typical reading would be 1×10^{-2} mbar, i.e. ten times higher. It must always be remembered when using a McLeod gauge that if any condensable vapours are present, the true total pressure will not be measured.

At high vacuum, penetration of mercury vapour into the vacuum system can be prevented by using a liquid nitrogen or similar cold trap fitted to the inlet of the gauge. This will also remove condensable vapours from the sample being measured and thus indicates the partial pressure of the non-condensable gases.

Occasionally, over a period of time, depending upon the application, the glassware may become contaminated with grease, oil, mercury scum, etc. Such contamination can lead to measurement errors. Chemical cleaning will be necessary following the maker's recommendations.

Typically the procedure will proceed as follows:

1. Remove grease with suitable solvent
2. Rinse out with tap water
3. Dissolve any mercury with nitric acid
4. Wash out with tap water
5. Clean the glassware with chromic acid
6. Drain and wash with tap water
7. Rinse with distilled water and dry

Appropriate care should be taken when handling mercury and acids, etc. Normally, if required, the scales, glass tubes, etc., may be replaced without recalibration because of the use of precision capillary tubes. This interchangeability can simplify gauge servicing.

A McLeod gauge is not normally found in industrial applications for reasons such as:

- It is fragile
- It is not able to measure condensable vapours
- Its output is not continuous (typically only one reading in two minutes is possible)
- It requires a skilled operator
- It is not adaptable to automation
- There is a danger of pollution by mercury

3.2 Transport phenomena gauges

Spinning rotor gauge

$$\text{Range: } 10^{-1} \text{ to } 10^{-7} \text{ mbar}$$

Pressure measurement can be achieved using the slowing down of a levitated ball-bearing caused by molecular drag effects between the ball surface and the gas. The slowing down depends on pressure, gas molecular weight and

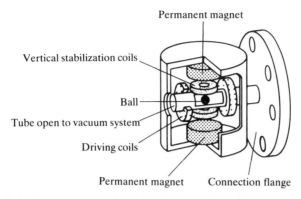

Figure 3.8 Spinning rotor gauge head partly exposed to show component parts

temperature of the gas, and on the surface state of the ball. It is a potentially very accurate gauge.

The gauge head consists of a steel ball, situated in a short tubular extension of the connection flange, as shown in Figure 3.8. This part is bakable up to 400 °C. The tube is surrounded by a magnet system for energizing and stabilizing the ball. Two permanent-magnet discs suspend the ball in the vertical direction. For stabilizing the ball in the centre of the magnetic field, a pair of control coils are coaxially arranged. For spinning the ball around its vertical axis a rotating field is produced by vertically arranged driving coils. When the ball has reached a given rotational speed of, say, 400 turns per second—i.e. 24 000 rev min^{-1}—further energizing is stopped and deceleration takes place, due to gas friction. The drop of rotational speed is measured as a function of time. Readings can be taken typically every 10 seconds.

The steel ball must have its own tiny magnetic dipole field for the gauge to work. It is important, therefore, that it is not overheated nor struck hard.

The gauge is being used as a vacuum transfer standard in several national institutions. Tests carried out in comparison with a series expansion system (see Section 3.6) have shown that the gauge reproduced readings over a two-year period within the limits of a total uncertainty of 1 per cent for the comparison with the standard.

Pirani and thermocouple gauges

Thermal conductivity gauges are electrically operated instruments which give continuous readings of total pressure in the range of 1013 to 10^{-4} mbar. They are reasonably robust and simple to operate and are very widely used. The physical principle employed is the pressure dependency of the ability of a gas to conduct heat.

In a typical instrument, the gauge head contains a fine-wire filament, usually tungsten, which is heated by passing an electric current through it (see Figure 3.9). The filament is exposed to the pressure to be measured. If the pressure is high, there will be frequent collisions between the gas molecules and the filament, and the molecules will take heat away from the filament and transfer it to the walls of the gauge head. If the pressure remains constant, the temperature of the filament will settle at some level. If the pressure in the gauge head is reduced, there will be fewer molecules present and therefore fewer collisions with the filament. This will mean that less heat is removed from the continuously heated filament, with the result that it settles at a higher temperature and the electrical resistance of the wire filament changes. The sensing filament in the gauge head forms one branch of a Wheatstone bridge and as the filament resistance changes the bridge becomes out of balance. The changing bridge current serves as a measure of the gas pressure. Note that since the voltage applied to the bridge is kept constant, this type of gauge is known as a 'constant-voltage' Pirani. Another type is the 'constant-temperature' Pirani; here the voltage applied to the bridge is regulated so that the resistance (and therefore the temperature) of the filament remains

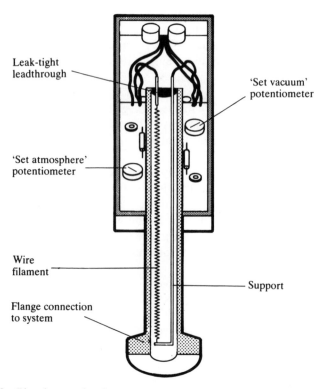

Figure 3.9 Pirani gauge head

constant independently of the heat loss. As the bridge is always in balance the voltage applied to the bridge is a measure of the pressure. Whereas the constant-voltage Pirani will measure pressure between 10 and 10^{-3} mbar, the constant-temperature Pirani has a larger measuring range from 1000 to 10^{-3} mbar. The gauges have very short response times and are particularly suitable for pressure-control applications.

Most Pirani gauges are calibrated for dry nitrogen; when other gases predominate, correction curves like those shown in Figure 3.10 must be used. In general, gases of low molecular weight have relatively high thermal conductivities at low pressure. Sensitivity is also dependent on the surface condition of the filament, i.e. whether it is contaminated. The presence of oil or other organic vapours can cause a high reading to be indicated. In general, cleaning of the filament with trichloroethane or acetone is possible but it must be done very carefully so as not to damage the sensor. The inside of the gauge head must be allowed to dry thoroughly before reconnection to the system.

When using a new gauge head with a control unit for the first time or when replacing a defective gauge head, it is usual to balance the bridge circuit. This is done by first checking the meter zero, i.e. for an analogue instrument, seeing that the meter pointer indicates atmospheric pressure. Gauge heads often have two potentiometers located within the gauge head but accessible internally. These enable the heads to be readjusted at 'zero' pressure (below

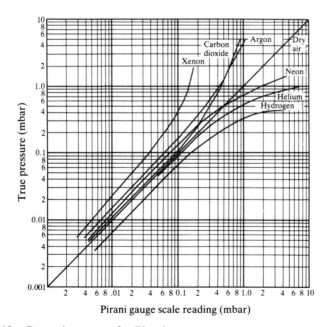

Figure 3.10 Correction curves for Pirani gauge

10^{-3} mbar) and at atmospheric pressure. Such potentiometers are usually marked 'Vac' and 'Atm' or alternatively '0' and '100%' (see Figure 3.9). Adjustment is made by inserting a screwdriver into the potentiometer.

The *thermocouple* gauge operates in a similar way to that of the Pirani except that the change of temperature of the filament is monitored by a thermocouple.

A thermocouple is an electrical thermometer working on a principle known as the Seeback effect. When two different metals are placed in contact, a small voltage appears across them. This voltage, which is at most a few millivolts, depends on the nature of the metals and on the temperature at the junction of the metals. In the thermocouple gauge head, the thermocouple junction is placed in close thermal contact (but not electrical contact) with the heated filament by embedding it in a tiny bead of glass (see Figure 3.11). The control unit consists of an electrical power supply for the filament, and a circuit and meter for measuring the output from the thermocouple.

The useful range of a thermocouple gauge is 10^{-3} to 5 mbar, with a best accuracy of approximately 10 per cent. Above 5 mbar the filament temperature changes very little. Typical filament temperatures are 450 °C at 10^{-3} mbar and 40 °C at 5 mbar.

This type of gauge uses very little power, permitting operation from a battery, allowing portability. To ensure that the filament current is set to the correct level, this instrument incorporates a 'set current' control which should be adjusted before the gauge is used.

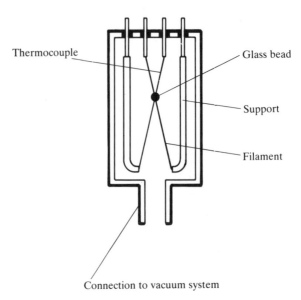

Connection to vacuum system

Figure 3.11   Section through a thermocouple gauge head

Neither the Pirani nor the thermocouple are harmed by sudden exposure to atmospheric pressure while working. However, care must be taken to ensure that any sudden inrush is not directed straight into the gauge head; otherwise the delicate sensing elements may be damaged.

Below 10^{-4} mbar, the thermal conductivity principle becomes insensitive. To measure high and ultra-high vacuum pressures another physical property of the gas is used, i.e. the measurement of ionization currents.

3.3 Ionization phenomena gauges

Cold cathode ionization (Penning) gauge

Range: 10^{-2} to 10^{-7} mbar

The principle of ionization is explained in Section 2.6. The simple device and circuit shown in Figure 2.4 could be (and in the past has been) used as a gauge—in two separate ways. When the device, known as a discharge tube, is operating, the ionized gas (plasma) will glow, and the shape of the glowing regions varies with pressure. It can give very rough indications of pressure in the range of 5 to 10^{-3} mbar. The second method of use is to observe the current flowing in the meter. Below 10^{-4} mbar, the current becomes very small.

The ionization is reduced to a very low level at this pressure because the mean free path of the electrons becomes very great, and they travel straight to the electrode and are lost, without colliding with any gas molecules.

In 1937 F. M. Penning, working for Philips in Holland, announced a modification of this type of gauge, incorporating a magnet, which overcame the problem. His gauge is variously known as the Philips gauge, the Penning gauge or the cold cathode ionization gauge. A modern gauge of this type (shown in Figure 3.12), known as a coaxial Penning, can measure pressures in the range from 10^{-2} to 10^{-7} mbar. The gauge consists of a straight-wire rod anode running through the centre of a cylindrical cathode. The magnetic field runs parallel to the anode. This geometry ensures that the magnetic field is everywhere perpendicular to the electric field, which is not necessarily the case in other designs of Penning.

When the gauge is switched on, several kilovolts are applied to the anode rod. There may be a short period of waiting for the ionization to build up. For the ionization to start, or 'strike', an ionizing particle, perhaps from natural background radioactivity, must appear inside the gauge and ionize a gas molecule. The waiting time is normally much less than one second but can be several seconds if the pressure is very low. It is therefore preferable to switch on the gauge at about 10^{-2} mbar. The ions and electrons are accelerated towards the cathode and anode respectively, and may ionize other molecules by collision. Further free electrons are produced when the ions bombard the cathode.

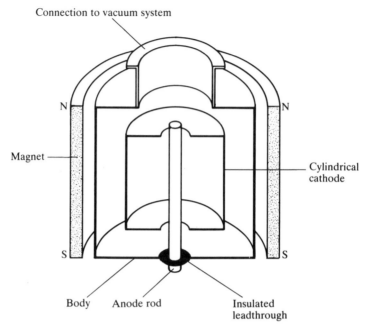

Figure 3.12 Section through a coaxial Penning head

These ions and electrons then produce other ions until a steady level is achieved where the rate of production of ions is equal to the rate of loss (either from arrival at the cathodes or from recombination with electrons). The ion current formed is displayed on a meter and is proportional to the pressure in the gauge head.

As the electrons accelerate towards the anode, their paths are bent into tight spirals by the magnetic field. This prevents them from immediately striking the anode to which they are attracted, and also greatly increases their path length.

Under ideal conditions, the accuracy of a Penning gauge is approximately $+100$, -50 per cent; i.e. a true pressure of 10^{-5} mbar might appear on the meter as anything between 5×10^{-6} and 2×10^{-5}. It is therefore only an indicator of pressure. For many applications this is all that is required.

It is recommended that the gauge should not be switched on until a pressure of 10^{-2} mbar has already been attained. After long operation at high pressure, sputtering occurs and a thin deposit of metal forms inside the gauge head. Although the deposits are themselves harmless they may eventually cause electrical leakage between electrodes.

During normal operation, contaminants such as oil vapours reaching the head can be decomposed ('cracked') causing carbonaceous deposits to form an insulating layer on the cathode, which may affect gauge head sensitivity and produce erroneous readings. Cleaning of the gauge head is reasonably simple, since the gauge head can normally be dismantled and the anode and cathode and other internal parts gently abraded with fine-grade glass paper to remove the contamination and produce a shiny surface on the anode. After abrasion all parts should be washed three or four times in a suitable solvent (e.g. acetone) and thoroughly dried. Cleaned parts should be assembled using gloves. Careful alignment of the magnet is often required on other types of Penning that are not of the coaxial design.

The Penning gauge is not a precision instrument but gives a rough indication of pressure.

Hot cathode ionization (ion) gauges

Range: 10^{-3} to 10^{-10} mbar

Penning gauges, while being rugged and simple to use, have poor accuracy and cannot measure pressures below 10^{-7} mbar. For greater accuracy or for lower pressures, hot cathode ionization gauges must be used.

The Penning gauge loses sensitivity below 10^{-7} mbar because the production of electrons for ionization falls to a very low level. This difficulty is overcome in the hot cathode gauge by fitting a heated filament (normally tungsten) which supplies free electrons at a constant controlled rate, ensuring that ionization will proceed at lower pressures. The principle by which the electrons are produced is known as 'thermionic emission'. In the bulk of any metal, a certain fraction of the atomic electrons are free to roam throughout the material (it is these that allow the metal to conduct electricity). The electrons are normally prevented from leaving the metal by a 'potential barrier' of a few volts which exists at the surface. If the metal is heated, sufficient energy can be given to some electrons to allow them to overcome the barrier and escape from the surface. These free electrons can then be drawn away by a nearby electrode at a positive voltage (the grid). Figure 3.13 shows a modern form of the gauge, known as a Bayard–Alpert type. The body of the gauge head is made of glass and glass-to-metal leadthroughs connect to the electrode structure. The grid has an open structure and many of the electrons pass right through and are then attracted back, oscillating back and forth several times before finally being captured by the grid. The electrons are thus given a long path length, increasing the probability of ionization. Any ions produced are attracted towards a third electrode—the collector (which in this design consists of a thin wire positioned in the centre of the grid). The resulting ionization current so formed is a measure of gas density and therefore pressure. The higher the pressure the greater the probability of an electron forming an ion.

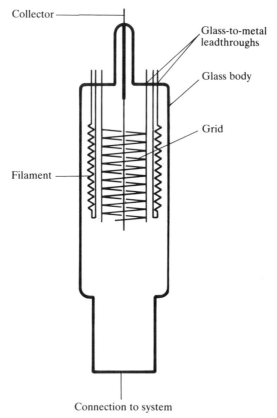

Collector

Glass-to-metal leadthroughs

Glass body

Grid

Filament

Connection to system

Figure 3.13 Hot cathode ionization gauge head

The ion current will also be proportional to the electron current (emission current) and it is usual to define the sensitivity (S) of an ion gauge by

$$S = \frac{I^+}{I^- P}$$

where P = pressure in mbar
I^+ = ion current
I^- = electron emission current (which is kept constant)

The gauge sensitivity, which is gas dependent, is determined by the geometric design of the gauge head. A value will normally be provided by the manufacturer. It is quoted in units of:

$$\frac{\text{amperes of ion current}}{\text{amperes of electron current}} \text{ per unit pressure}$$

and usually abbreviated to 'per mbar'.

The upper pressure limit of an ion gauge is typically 10^{-3} mbar. Above this pressure, the response becomes non-linear and there is a danger that the filament may burn out. Ion gauges commonly have a circuit which switches the filament off automatically if the pressure rises too high. The lower pressure limit is dictated by the electronic problems involved in measuring minute electric currents; e.g. for a gauge head reading 10^{-10} mbar with a sensitivity of 10 per mbar and an emission current of 1 mA, for nitrogen the current is of the order of 10^{-12} A.

A low pressure limitation is imposed by a 'false' pressure reading by X-ray emission from the grid. (Note that these are low-energy X-rays and are no danger to health.) As the electrons from the filament strike the grid, some of their energy is converted into X-rays. Many of these X-rays strike the collector and cause further electrons to be released from it. The loss of an electron from the collector is exactly equivalent to the collection of a positive ion as far as the measurement circuit is concerned. Thus even if the pressure is below 10^{-10} mbar, the X-ray limit due to bombardment causes the gauge to register a steady 10^{-10} mbar. An improved type of gauge known as a modulated ion gauge can measure pressures below 10^{-10} mbar.

As mentioned, the sensitivity is gas dependent, simply because the different gas molecules appear to the electrons as different sizes of target. The gauges are normally calibrated for nitrogen and any gas with a different 'cross-section' (as it is called) will cause an error in the reading. If the gas is of known composition, the correction factors shown in Table 3.1 can be applied to give an approximate true pressure reading for the Bayard–Alpert type of ionization gauge.

Several factors exist that may greatly affect the accuracy of the gauge. The composition of gases inside the gauge head is a particular source of uncertainty. At high or ultra-high vacuum very little of the gas present will be nitrogen—there are likely to be far greater amounts of water vapour, hydrogen or possibly oil vapours from the vacuum pumps.

The gauge itself can be a source of gas. Whenever the gauge is exposed to atmosphere, gases are sorbed on all the interior surfaces, to be released slowly

Table 3.1 Sensitivity of ion gauges to different gases and vapours

Gas	Factor	Gas	Factor
Acetone	4.0	Hydrogen	0.45
Argon	1.4	Mercury	3.6
Carbon dioxide	1.4	Neon	0.30
Carbon monoxide	1.1	Nitrogen	1.0
Helium	0.18	Water	1.1

The meter reading should be divided by the correction factor to give the true pressure reading.

when under vacuum. This will cause the gauge to read a falsely high pressure, particularly if it is connected to the system via narrow-bore tubing (a restriction will prevent the gases being pumped away quickly). The release of the gas from the surfaces can be hastened if the gauge power supply is fitted with a 'degas' facility. Operating the degas control causes the metal electrodes to be heated to 900 °C by electron bombardment from the filament. Degassing should always be carried out after exposing the vacuum chamber to atmospheric pressure, if accurate measurement is required. High-temperature baking of the gauge head should not be attempted at pressures above 10^{-5} mbar to avoid oxidation of the gauge elements, which can make it very difficult to attain ultra-high vacuum pressure readings.

Immediately after being degassed, the gauge head can act as a pump, since the very clean surfaces produced will trap gases until an equilibrium is established. Because of this, the pressure in the gauge can be lower than that of the system, again particularly so if the gauge is connected by narrow tubing.

Those latter two inaccuracies can be minimized by using a 'nude' type of gauge head—i.e. a gauge head mounted on a flange, rather than in a body, with the electrode assembly projecting right into the vacuum chamber. However, it has the disadvantage that the electrodes are susceptible to damage by activities in the chamber.

The high temperature of the tungsten filament can cause changes in the gas composition. A typical operating temperature is 1700 °C, and this is hot enough to break down many molecules into smaller fragments if they come into contact with it. Water vapour and hydrocarbons are likely sources. The effect can be reduced by using a filament made of rhenium and coated with lanthanum boride, which can emit sufficient electrons at the lower temperature of 1000 °C.

A common problem encountered with the use of ion gauges is the production of leakage currents through conducting layers inadvertently deposited on the inside of the gauge head, particularly in the region of the glass-to-metal leadthroughs. The deposits may be produced by evaporation of the filament. It may be possible to remove the deposits (particularly glass-bodied gauges) by electrical or chemical methods. Some examples of methods that have been suggested are detailed below. However, it is essential to refer to the manufacturer's recommendations before any such work is undertaken. Both methods have defects and the filament may have a special thoria coating that will rule out chemical cleaning. In such cases it is usually necessary to fit a new gauge head.

Note that with some coated filaments, in the presence of gases such as chlorine, iodine, bromine and fluorine or their compounds the coating can be worn off in a short time, causing the cathodes to burn through.

If cleaning methods are applicable they should be carried out with caution as befits the structure of the gauge head and personal safety. Complete removal of any cleaning compounds or chemicals is essential. It is beneficial to degas the gauge head in vacuum after cleaning.

Electrical cleaning
This method can be used after disconnecting the gauge head leads but without detaching the gauge head from the system. Thin conducting layers can normally be removed by connecting the ion collector to earth and applying a high-frequency potential from a spark tester unit such as that described in Section 14.4. This causes the film to 'burn out'.

Chemical cleaning
If the deposit is blue or silver grey in appearance it is likely to be an oxide of tungsten or molybdenum. These layers may be removed by soaking the gauge head in a 10 to 20% sodium carbonate solution, the temperature of which may be gradually raised to 30 °C to accelerate the reaction, taking precautions to avoid glass breakage. A pure film can usually be removed in about 15 minutes. If deposits are brown it is probably composed of cracked hydrocarbon vapour which should be removed with a 10 per cent solution of potassium hydroxide (allow to stand for half an hour). The electrodes may become discoloured by this treatment, but this will not affect the operation of the gauge head. Always conclude chemical treatments by rinsing thoroughly. Rinse the gauge head at least five times with tap water, then fill it with distilled water and allow it to soak for at least half an hour. Repeat the distilled water treatment twice more, ensuring that contact with the distilled water exceeds the period of contact with the chemical solutions.

3.4 Mounting of gauge heads (refer to Figure 3.14)

A vacuum gauge head should always be mounted as close as possible to the place where the pressure is required to be known. Inaccuracies can result if a gauge head is mounted at the end of a long pipeline, due to the resistance to gas flow along it. (This will be discussed in more detail in Chapter 13.)

Thermal conductivity gauge heads should be mounted vertically— otherwise the convection of gas around the filament will differ from the calibration situation and cause errors at higher pressures. All gauge heads should be mounted with the opening at the bottom, to prevent liquids or solid particles falling into them. Gas admittance through a valve into the vacuum chamber may cause a 'beaming' effect directly into the gauge head, giving rise to a false pressure reading. Since most gauge heads contain a number of fine-wire electrodes beaming might also damage the gauge structure. A more realistic indication will be obtained if the incoming gas stream can be directed and deflected from a system surface.

Pressure readings can additionally be dependent on gauge head position within the chamber. In systems being pumped or where there is a steady

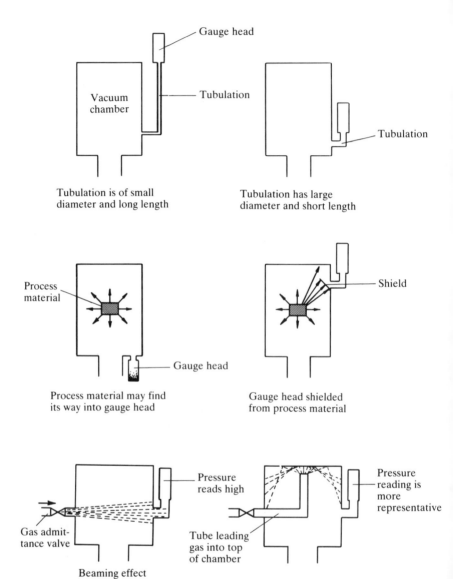

Figure 3.14 Mounting of gauge heads

transfer of gas from one region to another, generally the pressure will be lower nearer the vicinity of the pump inlet. The location of the gauge in relation to the pumps and to the region in which the pressure is required to be known should be taken into account.

When fitting several gauge heads to one component (tee or cross), they should be optically isolated from each other. Electrical and magnetic fields may cause measurement errors.

Do not leave gauge heads that are not connected to the system lying exposed on work surfaces. This applies particularly to Penning heads that incorporate magnets; they will attract dust and metal particles which can prevent good pump down times and cause erratic performance of the gauge.

3.5 Uncertainty surrounding the accuracy of measurement

We have already stated that the principal source of uncertainty surrounding the accuracy of most gauges arises from the lack of knowledge of the mixture of gases existing in the gauge head at any particular time.

All measurements are essentially inaccurate to some extent, so that what is called the 'actual' or 'true' reading is not attainable. It is normally possible to be able to estimate the limits between which the value being measured lies. The closer the limits, the more accurate the measurement. In most applications it is not always necessary to require a very high accuracy; a crude estimate may serve the purpose in that a process can be repeated successfully with good results.

Some typical examples of quoted gauge accuracies under ideal conditions are given below (note that accuracy is also often referred to as uncertainty):

1. Capsule dial $\pm 5\%$ of full-scale deflection
2. Capacitance manometer $\pm 1\%$ of reading or better
3. McLeod $\pm 10\%$ between 10^{-4} and 5×10^{-2} mbar
4. Spinning rotor $\pm 2\frac{1}{2}\%$ of measured value between 10^{-7} to 10^{-2} mbar and $2\frac{1}{2}$ to $13\frac{1}{2}\%$ between 10^{-2} to 1 mbar
5. Pirani $\pm 6\%$ between 10^{-2} and 10 mbar
6. Thermocouple $\pm 10\%$ between 10^{-2} and 1 mbar
7. Penning $+ 100$ to $- 50\%$, e.g. at 10^{-4} it can be 2×10^{-4} or 5×10^{-5} mbar
8. Ion gauge $\pm 10\%$ between 10^{-7} and 10^{-4} mbar
   $\pm 20\%$ at 10^{-3} and 10^{-9} mbar
   $\pm 100\%$ at 10^{-10} mbar

3.6 Calibration of vacuum gauges

In recent years there has been a considerable increase in the requirement for the routine calibration of vacuum measuring instruments, especially those used in manufacturing industries. The largest numbers of industrial processes are carried out in the pressure range from atmospheric to about 10^{-6} mbar so that the types of instrument most frequently requiring calibration are capsule gauges, Piranis, thermocouples and Pennings.

The reasons for the increased requirement for reliable day-by-day calibration may be because the quality of the end product is dependent on the quality of the vacuum produced so that reliable knowledge of the pressure in the system is essential. Often commercial contracts include a clause requiring that the product be manufactured to a specification in which the measuring instruments must be calibrated to certain specified levels of accuracy.

Much work has been carried out on the absolute calibration of vacuum gauges to the highest levels of accuracy. Calibration to accuracies of 1 per cent or less is restricted to a few national institutions which set up primary pressure standards and for that purpose use rather expensive and sophisticated equipment such as the two-series expansion system (indirect method) used by the National Physical Laboratory (NPL), England. These systems cover the ranges 10^{-8} to 10^{-1} mbar and 10^{-4} to 10 mbar. The estimated uncertainty in the generated pressure ranges from 1.6 per cent at 10^{-8} mbar to 0.46 per cent at 4 mbar.

The series expansion apparatus consists of a series of large and small volumes separated from each other by valves. The method allows an accurately known amount of gas contained in a small volume to be expanded into a larger calibration volume. If the temperature is maintained constant during the calibration process and the volume ratio of the vessels involved in gas expansion as well as the original pressure are known, the final pressure in the calibration chamber can be related to the original pressure (Boyle's law). Low pressures in the calibration chamber are obtained by repeating the process of gas expansion by up to four stages.

Use of the series expansion method requires an apparatus in which the base pressure must be low compared with the generated pressures (typically 5×10^{-10} mbar). The 'rate of rise' must be low also; the low-range apparatus at the NPL thus has to be a bakable stainless steel system using all-metal seals, but for the higher-range apparatus it has been possible to use elastomer seals in all the joints and valves.

Secondary standards (or transfer standard) gauges can then be calibrated against the primary standard. These secondary standards are commonly used by the NPL or by vacuum gauge manufacturers to calibrate reference standards (industrial working standard) for clients who may use this in turn to calibrate working gauges (see Figure 3.15).

For those involved in day-to-day calibration, absolute accuracy may not be important. Calibration by direct comparison with a secondary standard whose accuracy can be traced to a primary standard is generally sufficient. In such cases, the user will most likely use an industrial working standard in their own temperature-controlled calibration room using a simple direct

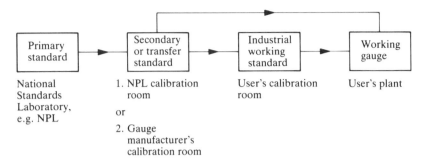

Figure 3.15 The different gauge standards produced by gauge calibration by comparison

comparison system (see Figure 3.16) to calibrate their working gauges. The system is based on a diffusion and rotary pump combination. The calibration chamber is to international standard recommendations and has six ports for the attachment of the appropriate master gauge and the gauge heads to be calibrated. Of particular importance is the manner in which the calibration gas is introduced into the chamber, the gas being reflected from the dome at

Figure 3.16 Calibration unit with associated electronic control console

the top of the chamber in order to produce a random distribution. The gauge head ports are arranged around the perimeter and with a right-angled bend ('elbow') in each connecting tube to ensure that each gauge sees exactly the same gas pressure. The system is not bakable since it is envisaged that the gauge heads to be calibrated will normally be joined to the system with fluoroelastomer 'O' rings. The calibration gas, normally dry nitrogen, is introduced into the chamber via a buffer volume.

Each of the master gauges is calibrated together as a matched pair of head and control unit and provided with a certificate of calibration giving levels of accuracy and showing direct traceability to NPL.

Calibration is thus by direct comparison comparing like with like, i.e. capsule gauge against capsule gauge and Pirani against Pirani gauge. The system would normally be set up in a room of its own or in a calibration laboratory and should only be operated by a technician specifically allocated to the job and trained accordingly.

There will be a need for the regular return of all master gauges to the calibration laboratory for rechecking and recertification. This is done typically yearly.

4

Identification of gases present

If, at any level of vacuum, information more detailed than a rough estimate of total pressure is required, then identification and individual quantification of every gas and vapour present becomes necessary. Instruments capable of this type of investigation, even at atmospheric pressure, are termed variously as partial pressure analysers, vacuum analysers, residual gas analysers or mass spectrometers. However, they all rely on maintaining high vacuum conditions internally for proper operation.

4.1 How mass spectrometers work

Mass spectrometers consist of four basic parts, shown in Figure 4.1:

1. An ionization stage where ions are produced (ion source)
2. A mass analyser stage where ions are separated according to their mass to charge ratio, so that only one type of ion proceeds to the
3. Ion detector, where the relative quantities of the ions are collected as a small ion current
4. The final stage consists of amplification and display of ion current for fixed emission current

These stages when combined together form an instrument that displays a spectrum of the masses of the molecules present and a measurement of the amount of each—hence the term mass spectrometer.

Two currently popular types of mass spectrometer will be covered in this section. Basically they differ in the way in which the ions are separated. Both utilize electromagnetic fields to cause the ions to follow different trajectories.

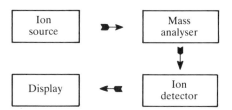

Figure 4.1   The four basic parts of a mass spectrometer

In one type a magnetic field is used; in the other an alternating electric field is used.

Magnetic deflection type (see Figure 4.2)

Electrons are emitted from the heated filament and attracted by a positive voltage on the electron collector. The electron beam passes through slits in the box-like ionizing region and ionizes any gas molecules that they encounter. The rate of electron emission is maintained constant, typically at 1 mA. The positive ions are attracted by a negative voltage towards the exit slit in the ion box. Some of the ions will strike the box and be lost. Others will pass through the slit, are accelerated and emerge as a beam of mixed ions travelling in a straight line. The beam of ions then comes under the influence of the magnetic field. (Usually a permanent magnet is mounted outside the mass spectrometer head.) This field drives the ions in circular paths of different radii (through typically 180°). Heavier ions tend to be deflected less than lighter ions. Hence the ion beam is separated into individual ion beams corresponding to each mass present within the sample. Each separate ion beam is normally brought to a focus at the ion collector by varying the accelerating voltage. Alternatively, variation of the magnetic field could be used; however, this type of control is more expensive. The ions, by transferring their charge to the collector, produce an ion current which is detected by the preamplifier and can be displayed.

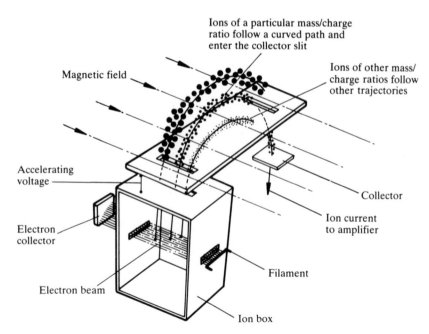

Figure 4.2 Magnetic analyser type mass spectrometer

Quadrupole type (see Figure 4.3)

Positive ions emerging from an ion source are focused into a beam and enter the quadrupole mass analyser. This analyser usually consists of a quadrant electrode assembly of four parallel circular rods which provide a specific radiofrequency field of a few megahertz. The rods are often located in

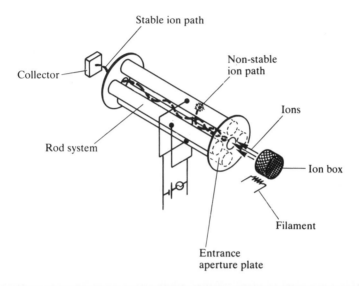

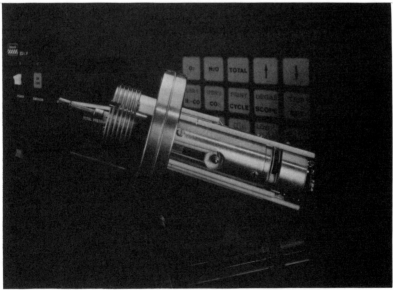

Figure 4.3 Quadrupole analyser type mass spectrometer. The photograph shows the analyser head with the control unit in the background

precision-ground alignment discs made of ceramic (not shown). In this form it is easy to dismantle the rods for cleaning and they are essentially self-aligning on reassembly. Opposite rods are electrically connected. The applied voltage consists of a constant d.c. component and a radiofrequency component. As a consequence of the oscillating field produced, positive ions entering the quadrupole region will oscillate. For a particular frequency only ions of a certain mass undergo stable oscillation and reach the collector. Those of higher or lower mass are collected on the rod electrodes. Mass selection is thus achieved.

Comparison of analyser type

Analyser	Principle
Magnetic	Uses motion of ions in a magnetic field
Quadrupole	Uses motion of ions in an applied r.f. and d.c. field for separation

Magnetic type
- Bulky
- Less sensitive to contamination than a quadrupole
- Magnetic field may interfere with other equipment
- Magnet may need removing for bakeout

Quadrupole type
- Compact
- Sensitive to contamination
- Does not use magnetic field
- May be cheaper than magnetic type

In general the quadrupole has found widespread acceptance as an instrument for gas analysis in vacuum systems, whereas the magnetic type is more often found in dedicated leak-detection equipment, which is specifically 'tuned' to look for helium search gas.

4.2 Some instrument characteristics

Pressure limit

The upper pressure limit for satisfactory operation is typically 10^{-4} mbar. At higher pressures the mean free path becomes significantly short, allowing ions to collide with molecules and become lost from the beam. Additionally, above 10^{-3} mbar there is a danger of oxidation and 'burning out' of the filament.

Sensitivity

The smallest detectable partial pressure is typically 10^{-12} mbar although some instruments with improved means of ion detection can be extended by a few orders of magnitude.

Mass range (for an explanation of atomic mass and a.m.u. see Appendix G)

This is a measure of the range of molecular weights that can be analysed. Some instruments are designed to measure only up to 80 a.m.u. Others have a range of several hundred atomic mass units. Generally small-sized heads have small mass ranges.

As we shall see, almost without exception, an upper limit of 100 a.m.u. is adequate to identify the gases found in the majority of vacuum systems. In fact for many applications there is no need to extend the analysis beyond 60 a.m.u.

Output facilities

In most modern instruments the partial pressures of the different gases can be displayed as a series of peaks or in a bar graph mode on a visual display unit. An output for a printer is also usually included. The instrument may have a series of preprogrammed channels which can be selected to give a digital display of partial pressures of some gases important in vacuum technology, e.g. argon, carbon dioxide, carbon monoxide, nitrogen, helium, hydrogen, oxygen and water.

Resolution (or resolving power)

This specifies the ability of the instrument to separate ions of neighbouring mass number.

4.3 Interpretation of mass spectra

Since mass spectrometers separate ions according to their mass-to-charge ratio, the resulting mass spectra have to be identified with the originating gas species. Ambiguity arises in cases where different molecules have the same mass.

This difficulty can be overcome by taking into account dissociation and multiple ionization of molecules, processes which are specific for different species. Dissociation occurs by chemical fragmentation during the ionization process and by thermal dissociation by the filament. Ions can have a multiple charge depending on whether they have lost one, two or three electrons during the ionization process. The ion is said to be singly, doubly or triply ionized respectively. Considering the hypothetical mass spectrum shown in Figure 4.4, the mass 28 peak is due to singly charged nitrogen ions. The molecular weight of a nitrogen atom is 14, but in the free state nitrogen is diatomic, i.e. each molecule consists of two atoms bonded together to give a molecular weight of 28 (N_2^+, where the '+' sign denotes a single charge). The 14 peak can therefore be due to doubly charged nitrogen molecules (N_2^{2+}) or singly charged nitrogen atoms (N^+). Similarly, the 32 peak is due to singly charged diatomic oxygen (O_2^+) and the 16 peak to the doubly charged molecule (O_2^{2+}) or singly charged oxygen atoms (O^+). Mass 18 indicates the presence of water vapour (H_2O^+) and mass 17 (OH^+) due to the breakdown of the water by electron impact.

The fractions of particles that dissociate or are doubly ionized under the conditions of mass spectrometry are fairly constant and therefore cause peaks of predictable size relative to the main peak. The resulting family of peaks is called the *cracking pattern*.

The commonest examples of gases with equal masses are N_2 and CO. Distinction between the two species is possible by analysing their cracking patterns. The parent molecule N_2 is mostly collected as N_2^+ at mass 28, a much smaller percentage being converted into N_2^{2+} and N^+, which appear at a mass-to-charge ratio of 14. The actual percentage number converted depends on the design of the instrument. The isotope of nitrogen (^{15}N) may also be recognizable as a very small peak. The molecule CO, on the other hand, produces several dissociation products such as C^+, O^+ and CO^{2+} which give rise to comparatively weak peaks at the masses 12, 16 and 14.

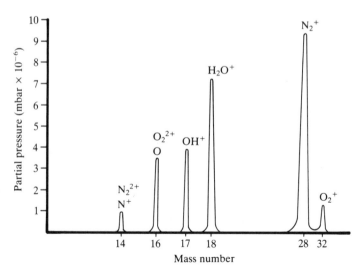

Figure 4.4   Hypothetical mass spectrum

Table 4.1 gives some typical cracking pattern data obtained in a magnetic sector instrument for some gases likely to be found in vacuum systems. Take, for example, the first column on the top line marked 'methane'. Moving down this column we reach a number '100.0', corresponding to mass 16. This means that for the methane spectrum the largest peak (taken as 100 per cent) appears at mass 16. Similarly, another peak at 15 would be 35.8 per cent of the 16 peak height. The 14 peak would appear as 18.6 per cent of the 16 peak height, and so on.

Composite mass peaks can be analysed by apportioning their total intensity to the different contributing species according to the intensities of the subsidiary peaks.

4.4 Qualitative interpretation of actual spectra—an example

A mass spectrum taken in a real system will almost always contain signals from a mixture of various gases. Although a quantitative evaluation is sometimes desired, one is often satisfied with a qualitative evaluation of the spectrum. A familiarity with some of the standard spectra of commonly expected gases will generally enable one to determine the major and minor gas phase components in the system.

An example of analysis in practice is shown in Figure 4.5. The system is a bakable ultra-high-vacuum system with a chamber of stainless steel construction. It is pumped by a rotary and diffusion pump combination.

Diagram (a)

Here with a maximum full-scale deflection (FSD) equivalent to 10^{-5} mbar, large nitrogen and oxygen peaks are seen, indicating an air leak into the system. The oxygen peak is equivalent to a partial pressure of about 2×10^{-6} mbar. Argon and water vapour are also discernable.

Action
The mass spectrometer was tuned to mass 4 (helium) and the outside of the system was probed for a leak using helium as a trace gas. A leak was detected and the problem rectified.

Diagram (b)

With the amplifier gain increased to 10^{-6} mbar FSD, a large mass 18 peak (partial pressure 9.2×10^{-7} mbar) due to water vapour is present. Adsorbed water vapour is tenaciously held to the system surfaces and comprises approximately 90 per cent of the total gas and vapour load at this level of vacuum.

Table 4.1 Mass spectra cracking patterns. (Information supplied by V. G.

	Methane	Ammonia	Water	Neon	Acetylene	Ethylene	Carbon monoxide	Nitrogen	Ethane	Nitric oxide	Methyl alcohol (methanol)	
Molecular weight	16	17	18	20	26	28	28	28	30	30	32	3
Formula	CH_4	NH_3	H_2O	Ne	C_2H_2	C_2H_4	CO	N_2	C_2H_6	NO	CH_4O	C
Mass												
2	3.0		0.7									
12	2.4				2.5	2.1	4.5					
13	7.7				5.6	3.5						
14	15.6	2.2			0.2	6.3	0.6	7.2	3.4	7.5		
15	85.8	7.5							4.6	2.4		
16	100.0	80.0	1.1				0.9			1.5		
17	1.2	100.0	23.0									
18		0.4	100.0								1.9	
19			0.1									
20			0.3	100.0								
21				0.3								
22				9.9								
24					5.6	3.7						
25					20.1	11.7			4.2			
26					100.0	62.3			23.0			
27					2.8	64.8			33.3			
28					0.2	100.0	100.0	100.0	100.0		6.4	
29						2.2	1.1	0.8	21.7		64.7	
30							0.2		26.2	100.0	0.8	
31										0.4	100.0	
32										0.2	66.7	1
33											1.0	
34												
35												
36												
37												
38												
39												
40												
41												
42												
43												
44												
45												
46												
47												

Quadrupoles Limited, Cheshire)

	Argon	Propene	Propane	Carbon dioxide	Nitrous oxide	Acetaldehyde	Ethyl alcohol (ethanol)	Nitrogen dioxide	Formic acid	N. butane	Acetone	Isopropyl alcohol
	40	42	44	44	44	44	46	46	46	58	58	60
₂S	Ar	C_3H_6	C_3H_8	CO_2	N_2O	C_2H_4O	C_2H_6O	NO_2	CH_2O_2	C_4H_{10}	C_3H_6O	C_3H_8O
0.2												
			0.4	6.0						3.3		
			0.5	0.1						2.9		
		3.9	2.5		12.9			9.6				
		5.9	3.9		0.1					5.3		
				8.5	5.0			22.3		5.2		
										17.1		
							5.5					
							2.3					6.6
10.7												
					1.2							
						1.6						
			0.7			4.8						
		11.3	7.6			9.1	8.3			6.3	8.3	
		38.4	37.9			4.5	23.9			37.1	8.0	15.7
			59.1	11.4	10.8	2.7	6.9		17.2	32.6		
			100.0	0.1	0.1	100.0	23.4		100.0	44.2	4.3	10.1
			2.1			31.1	6.0	100.0	1.6			
						0.1	100.0					
												5.6
4.4												
2.0												
0.0												
2.5												
4.2	0.3		0.4									
		13.4	3.1								2.1	
	0.1	20.3	4.9								2.3	
		74.0	16.2							12.5	3.0	5.7
	100.0	29.9	2.8									
		100.0	12.4			3.9				27.8	2.1	6.6
		69.6	5.1			9.2	2.9			12.2	7.0	4.0
			22.3			26.7	7.6			100.0	100.0	16.6
		26.2	100.0	100.0	45.7					10.0		
		0.8	1.3	0.7			34.4		47.6			100.0
		0.4	0.2				16.5	37.0	60.9			
										(12.3 at 58)	(27.1 at 58)	(3.4 at 59)

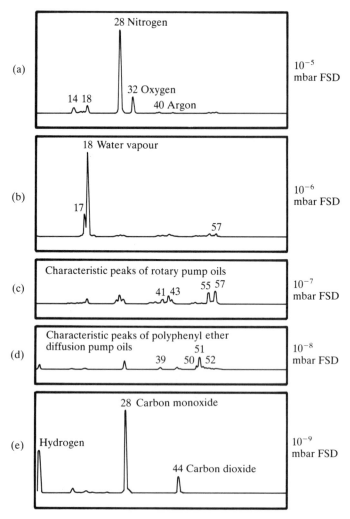

Figure 4.5 An example of analysis in practice. (Reproduced by permission of V. G. Quadrupoles Limited, Cheshire)

Action

The system was baked ensuring no cold spots existed within the system for desorbed water vapour to transfer to.

Diagram (c)

With the amplifier gain increased to 10^{-7} mbar FSD, high partial pressures of masses 41, 42, 43, 55 (1.1×10^{-8} mbar) and 57 (1.4×10^{-9} mbar) indicate hydrocarbon from rotary pump oil back-migration.

Action

An activated alumina foreline trap was fitted to the rotary pump (see Section 5.6).

Diagram (d)

With the amplifier gain increased to 10^{-8} mbar FSD, peaks in the mass range 50 to 54 are noted with mass 51 of partial pressure 1.4×10^{-9} mbar. These are characteristic of polyphenyl ether diffusion pump fluid.

Action

A liquid nitrogen trap was fitted above the diffusion pump.

Diagram (e)

With the amplifier gain increased to 10^{-9} mbar FSD, three major peaks are seen: hydrogen (5.4×10^{-10} mbar), carbon monoxide (9.8×10^{-10} mbar) and carbon dioxide. The source of these gases is explained in Section 4.5, but basically they originate from the materials within the vacuum system.

4.5 Sources of residual gas

We have seen that the gas composition in the vacuum chamber will depend on the type of pump used, the choice of working fluid (if any), the design and construction of the chamber (metal or fluoroelastomer seals, for example), whether the system has been baked, etc. To assist in the interpretation of spectra, Table 4.2 lists possible sources of some residual gases left in the vacuum system.

A knowledge of the recent history of the vacuum system may provide a clue as to the possible or probable gases that may be present in the vacuum system. Some points worth considering are:

1. Has the system been exposed to the atmosphere?
2. Could a leak have developed as a result of a system modification being made?
3. If a modification has been made were any high vapour pressure components used in its construction?
4. Has the system been baked out? How well was it baked?
5. Did any solid compounds decompose or react in the system?
6. Were there reactive gases in the system?
7. Has the system been exposed to hydrocarbon contamination?
8. Who used the system recently? What did they do? ... And so on.

Table 4.2

Peak mass (a.m.u.)	Gas	Source
2	Hydrogen	Hydrogen is normally the dominant species in UHV systems. It can be the result of desorption from the materials of construction of the vacuum system (e.g. stainless steel). This is slowly and continuously released. It can also originate from the dissociation of water vapour or pump fluids.
4	Helium	Helium will permeate through fluoroelastomer seals and Pyrex at UHV. It can also be residual helium from leak testing remaining in fluoro-elastomers for some time.
16	Oxygen O^+	See notes for fluorine (19), etc.
16	Methane CH_4	Synthesis within systems. An active titanium surface in ion pumped and titanium sublimation pumped systems seems to enhance the synthesis. Methane has peaks at 14 (CH_2^+), 15 (CH_3^+) and 16 (CH_4^+).
		Mass no: 13 14 15 16
		Abundance (%): 8 16 86 100
17	Ammonia NH_3	Characteristic peaks at 17, 16, 15 and 14. In UHV systems ingressing N_2 from a small air leak can react with residual H_2 to form NH_3.
18	Water vapour	Water vapour will dominate residual gas composition of unbaked systems. During a bakeout, especially at the start, water vapour is released from the system walls very rapidly. Characterized by a family of peaks at 16 (O^+), 17 (OH^+) and 18 (H_2O^+). The water vapour cracking pattern is typically:
		Mass no: 16 17 18
		Abundance (%): 1 23 100
19 23 35, 37 39	Fluorine Sodium Chlorine Potassium	Fluorine can be desorbed from stainless steel surfaces within the mass spectrometer ion source due to electron bombardment of the surfaces. Similarly for sodium, chlorine and potassium; this effect is often accompanied by mass 16 (O^+).
20	Neon	20 and 22 peaks are neon and its isotope (ratio is approximately 10 to 1). Often seen in UHV ion pump systems, since neon is not pumped very well. A large 22 peak may be CO_2^{2+}.
28	Carbon monoxide	Carbon monoxide is often produced by carbon in the tungsten filament of the ion source as an impurity or by earlier contamination of the filament with system hydrocarbons, forming a carbide which subsequently reacts with residual system oxygen or water vapour to form CO. Ion gauges are often turned off in UHV to reduce the CO background level. CO is commonly the second major constituent in a system.

Table 4.2 (continued)

Peak mass (a.m.u.)	Gas	Source
28	Nitrogen	Probable air leak. For N_2 the N^+ peak is about 7 per cent of the N_2^+ peak height. The presence of 32 and 18 would confirm a leak.
30	Nitrogen oxide	NO^+ may be formed by N_2 and O_2 reaction at hot filament.
30	C_2H_6	Hydrocarbon contamination.
32	Oxygen	A large 32 peak in association with a higher 28 peak can indicate an air leak.
35, 36, 37, 38	Chlorinated compounds	Possibly from cleaning residuals. Can also be produced by electron bombardment of some surfaces.
36 38	$H^{35}Cl^+$ $\}$ $H^{37}Cl^+$	Possibly due to cleaning fluids or the reaction of H and Cl.
40	Argon	Probably an air leak. A 20 peak about 10 per cent as large as the 40 peak would be further evidence of an air leak.
44	Carbon dioxide	Generated by ion gauges in a similar manner described for CO. This is common shortly after turning on filaments. High CO also present would confirm. Also due to desorption from internal surfaces, particularly unbaked systems. A peak at 22 (CO_2^{2+}) is 1/10 peak at 44 (CO_2^+).
16 30 44 58	CH_4^+ $\}$ $C_2H_6^+$ $C_3H_8^+$ $C_4H_{10}^+$	Hydrocarbon contamination with clusters at an average interval of 14 a.m.u. apart, corresponds to the removal of one or more (CH_2^+) groups during ionization.
41 43	$C_3H_5^+$ $\}$ $C_3H_7^+$	Hydrocarbon contamination typical of the degradation of some hydrocarbon plastic-like materials.
41, 42, 43, 55, 57 $\}$		Probably mechanical pump oil Cracking pattern (major peaks)

Mass no: 41   43   55   57
Abundance (%): 33   73   73   100

| 45 | | Isopropyl alcohol
Cracking pattern (major peaks) |

Mass no: 27   29   43   45
Abundance (%): 16   10   17   100

| 58 | | Acetone (possible remains of cleaning fluid)
Cracking pattern (major peaks) |

Mass no: 43   58
Abundance (%):   100 27

| 78 | | Trimethyl pentaphenyl trisiloxane
Silicone oil (DC or MS 705)
Cracking pattern (major peaks) |

Mass no: 39   43   76   78
Abundance (%): 73   59   83   100

(continued)

Table 4.2 (*continued*)

Peak mass (a.m.u.)	Gas	Source
95		Trichloroethylene (cleaning fluid)

95 — Trichloroethylene (cleaning fluid)
Cracking pattern (major peaks)

Mass no:	60	95	97	130	132
Abundance (%):	65	100	64	90	85

97 — Trichloroethane (fluid used for degreasing components)
Cracking pattern (major peaks)

Mass no:	26	61	97	99
Abundance (%):	31	58	100	64

446 — Polyphenyl ether (diffusion pump fluid)
Cracking pattern (major peaks)

Mass no:	39	51	77	446
Abundance (%):	10	28	79	100

4.6 Differential pumping

In the majority of applications for which mass spectrometry is used, the total pressure is low enough that the mass spectrometer can be connected directly to the vacuum chamber. The upper limit of this pressure is about 10^{-4} mbar.

Some processes, however, such as plasma etching or sputter deposition, operate at higher pressures up to several millibars. In such cases a differentially pumped configuration may be exploited. In this method gas from the high-pressure system leaks into the high vacuum of another continuously pumped subsidiary vessel to which the mass spectrometer is attached. Leakage through a sintered porous disc is a typical example of the way in which this is achieved. Thus the process gases can be characterized by the relative peak heights in the fingerprint spectrum, and deviations from normal values may be revealed in the standard fashion.

4.7 Applications of mass spectrometry

Some examples where mass spectrometry can be used:

- In the analysis of process gases, perhaps for quality control purposes
- In fault finding in a vacuum system, to distinguish between real leaks and other system problems
- In the analysis of outgassing of materials and components in vacuum systems
- In hospital operating theatres for respiratory gas analysis during anaesthetization

Oil-sealed mechanical rotary pumps

Two of the most commonly used designs of this type of pump are the rotary-vane and rotary-piston pumps.

5.1 Rotary-vane pump

This type of pump is the one most generally adopted in the vacuum industry today for the production of pressures down to 10^{-2} to 10^{-3} mbar. Figure 5.1 illustrates, in schematic form, the cross-section of a typical rotary-vane pump. The mechanism, which is lubricated by oil, consists of a housing (stator) with

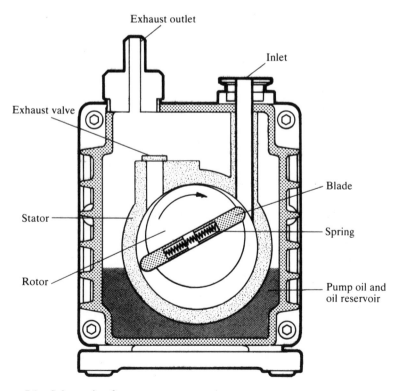

Figure 5.1 Schematic of a rotary-vane pump

a cylindrical bore into which is fitted a rotor. The rotor is offset in relation to the stator bore and fits closely against the stator in one position. The stator bore in this area has a curvature equal to that of the rotor accurately machined across the whole width of the stator bore. The rotor contains two blades (generally spring loaded) which slide in diametrically opposed slots. Thus, as the rotor turns, the tips of the blades are in contact with the stator wall at all times.

Figure 5.2 shows four stages in one revolution of the rotor. The cycle is divided into induction, isolation, compression and exhaust phases. During operation gas molecules entering the inlet of the pump pass into the volume created by the eccentric mounting of the rotor in the stator. The crescent-shaped gas volume is then compressed, forcing the exhaust valve open and

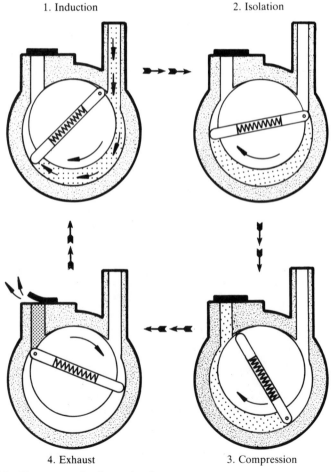

1. Induction   2. Isolation

4. Exhaust   3. Compression

Figure 5.2   Four stages in the cycle of a rotary-vane pump

permitting gas discharge. Note that there are two pumping 'cycles' per revolution. In order to produce a seal which will combat the very high pressure difference between the inlet and discharge sections, the gap between the rotor and stator is required to be of a very small order (typically 0.025 mm (0.001 in)).

To complete the seal, a thin oil film is continuously maintained between the components by oil drawn from the oil reservoir into the pump interior. The seals necessary between the blade edges and tips and the stator are made in the same way. The circulating oil is ejected back into the reservoir through the exhaust valve together with the pumped gas. The ultimate pressure is limited by back-leakage between the suction and discharge compartments and by outgassing of the lubricating oil.

In another version of this pump, improved performance is achieved by using two stages in series to produce a two-stage pump. The first or high vacuum stage, shown in Figure 5.3, is 'backed' by the second or low vacuum stage via an internal transfer duct. With this arrangement the pressure at the exhaust port of the high vacuum stage is considerably less than atmospheric pressure when the inlet pressure is low. Thus, the pressure difference, and hence leakage across the sealing regions of the high vacuum stage, is much reduced, enabling lower pressures to be achieved before leakage begins to affect the performance. Oil for lubricating and sealing the high vacuum stage

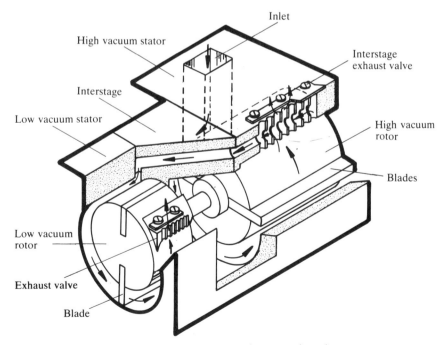

Figure 5.3   Two-stage rotary-vane pump, partly exposed to show component parts

is degassed by the low vacuum stage before being passed to the high vacuum stage. This, together with the sharing of compression between two stages, produces a significantly lower pressure than a single-stage pump. The pumping volume of the high vacuum stage is often considerably larger than that of the low vacuum stage (the low vacuum stage components may be used for a number of different pump sizes). In pumps with different capacities of high and low vacuum stages, a pressure relief is provided between the stages so that there is no overpressure in the interstage duct which would otherwise occur when the pump is operating at, or just below, atmospheric pressure. The valve automatically shuts when the interstage pressure drops below atmospheric pressure.

One advantage of the rotary-vane design is the comparatively small out-of-balance forces consequent upon rotation. The only eccentrically rotating masses are the blades, the centre of mass of a blade pair describing an oval path twice per rotor revolution. With the use of various modern plastic blade materials, which are somewhat lighter than steel, this imbalance becomes insignificant. The use of steel blades in the older type of pumps presented a limit on size and rotational speed of the pump; excessive peripheral blade speed led to severe heating by friction, leading sometimes to seizure. With modern plastic blade materials peripheral speed can be increased. This has eased the problems associated with directly driven pumps.

Most rotary pumps used to be belt-driven, having rotational speeds usually of between 350 and 750 rev min^{-1}. Modern trends in design towards compact plant have led to pumps being directly coupled to their motors to run at the speed of a four-pole motor ($\simeq 1400$ rev min^{-1} on 50 Hz mains). However, higher running speeds have accentuated the noise problem. Noise originates from two main sources, namely hydraulic knocking and noisy exhaust valves. Hydraulic knocking is associated with the quantity of sealing oil entering a pump; the amount is critical for production of optimum vacuum conditions and must be carefully controlled.

Too much oil will produce hydraulic knocking and outgassing of the oil excess will adversely affect the ultimate pressure, while too little oil will not fulfil the sealing and lubricating requirements. A positive means of metering the required amount of oil into the pump is thus required.

Early designs of pump (where the whole assembly is immersed in oil) permitted oil entry during exhaust while the valve was open. One early type of exhaust valve manufactured between the late thirties and early fifties consisted of a steel ball moving in a coned seating restricted in upward movement. Another type was a flat metal strip over an exhaust gas hole with side-locating pillars having a bridge across them with an adjuster for restricted movement. The problems with these types of exhaust valve were that they were extremely noisy (it was claimed that an estimate of the quality of the vacuum could be made by listening to the varying sound characteristics!), that any alteration of exhaust valve lift due to wear would change the quantity of oil entering the pump and that pump stoppage with pressure difference existing across it caused atmospheric pressure to push oil into it and then up into the vacuum system, causing contamination—this effect was known as 'suck-back'.

Many procedures to prevent or reduce 'suck-back' have been made with various types of electric, manual and float valves inserted between the pump and vacuum system, all having limited success. A more definite 'suck-back' preventer was required; provision for a positive supply of oil to the pump and a reduction of exhaust valve noise was also needed.

A synthetic rubber type of flap exhaust was introduced in the mid-fifties; this covered the row of exhaust gas holes and was backed by a metal plate to limit lift and ensure return. The flap lies in a depression on top of the pump body containing only a little oil. On pump stoppage atmospheric pressure flattens the flap onto the exhaust holes, sealing against oil entry except where foreign bodies prevent the flap seating. Even then, the small amount of oil in the depression is insufficient to cause trouble. This type of valve is very quiet.

One method of supplying oil to the pump is from a positive pressure oil

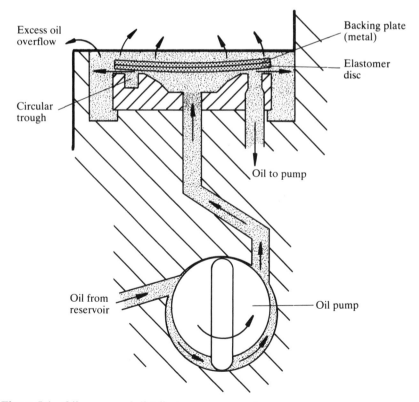

Figure 5.4 Oil pump and distributor arrangement

pump, built into the pump shaft, via a distribution arrangement shown in Figure 5.4. The distributor sits in an oil-filled depression similar to the exhaust valve already described. The oil pump develops a positive oil feed pressure of typically 0.4 bar above atmospheric pressure in the supply line to the distributor. This pressure is sufficient to lift and hold the elastomer disc away from its sealing face and allows the oil to overflow into the circular trough. From the trough the oil passes to the pump interior to fulfil its necessary functions and eventually it finds its way back to the reservoir. When the pump is switched off, the pressure in the oil feed line is no longer maintained and the elastomer disc, which is reinforced with a sprung backing plate, moves down, sealing off the exit from the distributor to the pump interior. The pump is then effectively isolated and does not suck back oil.

In the event of either the distributor or the exhaust valve failing to seal, it will be air that is sucked in, not oil.

This arrangement, which eliminates any quantity of oil entering the pump, also eliminates any difficulties in starting the pump after some hours of rest. Previously, oil found its way into the pump stator and caused the pump to lock hydraulically on start-up, in some instances necessitating the turning of the pump by hand to clear the oil. The rotary-vane pump has thus changed from a simple but noisy pump with 'suck-back' problems to a quiet and reliable unit which is now directly driven.

5.2 Rotary-piston pump

The rotary-vane pump design has been limited to relatively small capacity pumps since there was a limit on the peripheral speed of steel blades due to friction and consequent oil breakdown. In general, larger capacity pumps were of a different design, i.e. rotary piston. With the advent of plastic-reinforced blades this situation is now changing. Figure 5.5 shows a rotary-piston type of pump where the piston (a hollow cylinder with a hollow tongue attached) is made to precess around inside a circular stator by means of the rotating cam within it. The operating cycle, i.e. increase volume, isolate, compress and discharge, are similar to the basic movements described in the rotary-vane pump. These pumps are generally belt-driven and the inevitable out-of-balance forces that exist can be partially balanced by a counterweight on the driving pulley.

5.3 Gas ballasting

When pumping gas loads containing vapours such as water vapour, there is a possibility that condensation of the vapour may occur, should it reach its saturated vapour pressure during pump compression. Liquids thus formed can mix with the pump oil—water, for example, forms an emulsion. Such a condition can rapidly lead to:

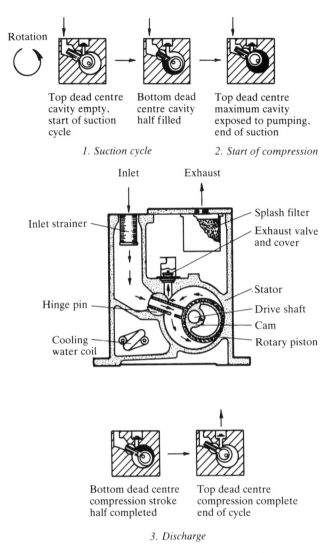

Figure 5.5 Rotary-piston pump and stages in operation of the pump

1. Deterioration of vacuum performance
2. Reduction in the lubrication and sealing properties of the oil
3. Eventual corrosion of the pump components

Two possible solutions to this problem would be to:

1. Change the oil at regular intervals (expensive and inconvenient).
2. Trap the water vapour prior to its entry into the pump by using a

desiccant trap, a water-cooled condenser, a mechanically refrigerated condenser or a liquid nitrogen trap.

However, most modern pumps are fitted with a 'gas ballast' facility, which enables the pump to cope with a certain amount of vapour without contamination of the pump oil. This device is usually visible as a control valve mounted externally on the pump body.

The gas ballast facility allows atmospheric air (a dry or inert gas can alternatively be used) to be admitted to the pump during the compression stage. This increases the proportion of non-condensable gas in the pump. Consequently, the partial pressure of the vapour being pumped at the time the exhaust valve lifts does not now exceed its saturated vapour pressure at pump temperature. The vapour is thus discharged without condensing. A simple non-return valve is usually incorporated into the gas ballast inlet which ensures that the pump exhaust does not leave through the ballast inlet.

Due to the extra work involved in compressing the additional air admitted to the pump, the pump temperature rises. This temperature rise helps to prevent vapour condensing inside the pump.

Running on gas ballast affects the ultimate pressure, especially on single-stage pumps, due to the increased leakage across the seal between the high- and low-pressure sides of the pump. A single-stage pump may not give an ultimate pressure of better than 0.5 mbar on gas ballast (compared with 5×10^{-3} mbar without gas ballast). Note that 0.5 mbar may not be sufficient for 'backing' a diffusion pump. On two-stage pumps the effect is less since only the low-vacuum stage of the pump is gas ballasted. Ultimate pressure of the order of 5×10^{-3} mbar can be expected with gas ballast (compared with 5×10^{-4} mbar without gas ballast).

Note that the values of ultimate pressure quoted are partial pressures of the permanent gases present and thus ignore any contribution due to oil vapour that may be present in the vicinity of the pump inlet. Measurements using total pressure gauges (e.g. the Pirani gauge) will obviously give higher readings, and these would be dependent on the vapour pressure of the oil (hence also pump temperature) and possibly on the oil contamination level.

It may be useful at the beginning of a pumping process to run with the gas ballast valve open. On the walls of a vacuum chamber there is almost always a thin film of water (derived from the moisture in the air), which vaporizes gradually at first. When the majority of this vapour has been removed the gas ballast valve should be closed, in order to obtain a low ultimate pressure. Note that not all of the water vapour will be removed; in fact, in unbaked high vacuum systems adsorbed water vapour comprises approximately 90 per cent of the total gas load.

Another benefit of gas ballast is that it can be used on contaminated pumps to purge the pump oil of condensed vapour (see Table 5.1). In this case ballasting should be done for a minimum of 2 hours to allow the pump to

Table 5.1   Gas ballast guidelines

Condition of system	Gas ballast condition
System containing atmospheric air, regularly cycled	If possible turn the gas ballast on *only* during the roughing period. If this is not possible, gas-ballast continuously but with the control valve only partially open, or at least gas-ballast the pump at regular intervals with the control valve fully open.
Pumping system containing large amounts of vapour but below maximum recommended vapour pumping rate	Gas-ballast continuously with the control valve fully open.

Condition of pump	Gas ballast condition
Reclaiming contaminated oil (oil/ vapour emulsion)	Full gas ballast isolated pump for at least 2 hours
New oil in pump (pump cold)	ditto for 1 hour
New oil in pump (pump hot)	ditto for $\frac{1}{2}$ hour

thoroughly warm up. After this period of time, and after the gas ballast is shut, a noticeable improvement in the ultimate pressure can be demonstrated.

If an application demands that a pump must run on continuous gas ballast care must be taken to ensure that the pump does not run short of oil, since under these conditions oil mist is continuously discharged at the outlet. Additionally, the exhaust from the pump, if piped away, requires that the pipeline be taken from the pump initially in a downward direction, or preferably into a suitable condensate trap. Otherwise, if the exhaust line leads directly upwards from the pump any vapour condensing in the line will fall back into the pump oil reservoir, possibly causing oil contamination to occur.

Even with gas ballast there is a limit to the quantity of vapour that a pump can handle and values are normally available for each pump, e.g.

For a 80 $m^3 h^{-1}$ rotary-vane pump	Maximum water vapour pumping rate (kg h^{-1})
Single stage	1.2
Two stage	0.3

In general this is equivalent to a water vapour pressure at the inlet of 30 mbar for the single-stage pump and 5 mbar for the two-stage pump.

The maximum amount of vapour that can be handled by any pump is determined by the gas ballast air flow rate. In most modern two-stage pumps only the low vacuum stage is gas ballasted (because it is unlikely for condensation to occur in the high vacuum stage). Since the low vacuum stage is usually of smaller capacity than the high vacuum stage, gas ballast flow rates are less for two-stage pumps than for single-stage pumps.

Note that to handle greater vapour pumping rates than those quoted would require some form of vapour trapping of the gas load prior to entry into the pump.

The following points should be noted concerning gas ballast:

- Purging contaminated oil takes time—2 hours plus
- Exhaust lines—the condensate must not drain into the pump
- Gas ballasting spoils the ultimate
- With badly contaminated systems, let pump warm up by running for some time before evacuating the system
- An increase of pump temperature with ballasting improves vapour pumping capacity

5.4 Rotary pump oils

The functions of oil in a rotary pump are:

- Sealing
- Lubrication
- Cooling
- Corrosion protection

The types of oil employed for oil-sealed rotary pumps have vapour pressures at least as low as 10^{-4} mbar at room temperature. They are generally mineral oils refined to eliminate, to varying degrees, higher vapour pressure constituents. When in use the ultimate pressure can fall initially as the fluid is purified by distillation, but this is opposed by friction effects at rubbing surfaces which promote fluid degradation. Rotary pump oils supplied by various manufacturers are very similar in that they offer low vapour pressures even at pump maximum operational temperatures. Their water adsorption rate is minimal and they have good lubricating qualities.

The requirements for rotary pump oils are

- Sufficiently low vapour pressure
- Suitable viscosity
- Adequate lubricating properties
- Chemically suitable in relation to materials being pumped
- Chemically stable at pump working temperatures
- Not hazardous to health

Originally the only oils available for rotary pumps were purified mineral oils. For some applications these had additives included to reduce, for example, oxidation, corrosion and foaming. However, for some uses mineral oils are far from satisfactory; the pumping of oxygen or acids is a prime example. Alternative synthetic fluids are now available for such difficult applications.

Mineral oil lubricants

Mineral oils are the most commonly used fluids for rotary pumps. In general, oils available from a garage/service station or store do not have a suitable low vapour pressure and none exhibit the service life or low level of back-migration required for most processes. Mineral oils offered by vacuum pump manufacturers and other suppliers have been distilled to remove volatile components.

Singly distilled fluids
Singly distilled fluids are suitable for non-corrosive applications and have a longer life (five times the service life) and lower vapour pressures than their undistilled counterparts. These fluids are available in a range of viscosities:

1. Low-viscosity fluid is normally used in low-temperature conditions where starting the pump may be difficult. Additionally, it may be available with antioxidant additives. This is useful for operating conditions that generate unusually high temperatures, such as in vacuum packaging applications.
2. Medium-viscosity fluid is normally suitable for piston pumps and direct-drive rotary-vane pumps. This fluid is also available with multifunctional additives to give protection against corrosion. The presence of additives increases the vapour pressure so that the ultimate pressure obtainable is slightly worse than that obtained without additives.
3. High-viscosity fluid is normally used for belt-drive rotary-vane pumps which can provide slightly higher starting torques due to the gearing of the belt pulleys. It has the benefit of a slightly lower vapour pressure than the medium-viscosity fluid. This fluid is also available with additives.

Figure 5.6 shows how vapour pressure varies with temperature for typical mineral oils used in rotary pumps. Note that operating temperatures of rotary pumps are typically in the region 80 to 90 °C.

Doubly distilled fluids
Doubly distilled fluids have much lower vapour pressures than singly distilled varieties. The chief advantages are lower rates of back-migration and a longer service life.

White oil
Technical white (TW) oil is a hydrocarbon fluid which is distilled from a stock that has been extensively processed to remove naturally occurring

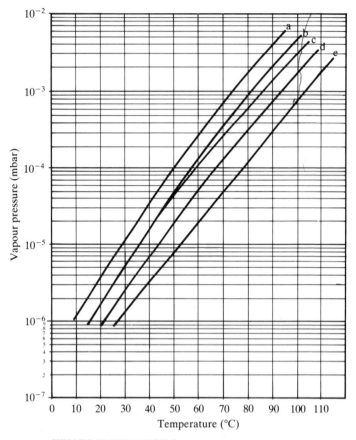

KEY TO TYPES OF OILS
a Medium viscosity with additives
b Low viscosity
c Medium viscosity
d High viscosity with additives
e High viscosity

Figure 5.6 Vapour pressure/temperature curves of typical rotary pump oils. (By kind permission of Shell Lubricants UK, Cheshire)

impurities. As a result, these fluids have service lives two to three times longer than doubly distilled fluid in comparable applications. It is therefore recommended for applications where exposure to reactive or corrosive gases is prevalent.

Inert mechanical pump fluids

When pumping oxygen, mineral oil can react explosively; therefore a fire-resistant fluid must be used in the rotary pump. Such fluids as phosphate esters or proprietary brands such as Fyrquel* can be used. However, these fluids attack most elastomers and therefore fluoroelastomers seals and exhaust valves should be used.

Inert mechanical fluids (such as perfluoropolyether—PFPE) are used because of their total oxygen compatibility and their stability when pumping other reactive/corrosive gases. Proprietary brands of this fluid such as Fomblin† and Krytox‡ when used in rotary pumps with suitable accessories make them of unique value, e.g. in most semiconductor process applications (see Section 5.8).

In changing from mineral oil to PFPE fluids it is necessary to clean the pump as thoroughly as possible before charging with the replacement fluid. The degree of cleanliness depends on the application. For oxygen pumping, *complete* strip down, thorough cleaning and reassembly of the pump is *essential.* Consideration should also be given to renewing some of the non-metallic parts, such as seals and blades. For other aggressive applications, complete strip down is also recommended.

During strip down, mineral oils can be removed with a solvent such as trichloroethane. If it is required to remove PFPE from a pump, a fluorinated solvent such as trichloro-trifluoroethane must be used for the cleaning process, e.g. Arklone P, Algofrene 113, Arcton 113, Freon 113. It is important that oil be drained from a pump in a safe manner depending upon the application. For example, analysis of pump oil from a semiconductor etcher using carbon tetrachloride has revealed the presence of trichloroethane, perchloroethane, toluene, etc. None of these is safe to breathe, in large quantities, or to ingest through the skin.

Table 5.2 lists some typical applications for some of the fluids described, together with some of their physical properties. Table 5.3 lists some typical oil price comparisons.

5.5 Pumping speed and ultimate pressure

Pump speeds for rotary pumps are often quoted as the volumetric displacement of the pump at atmospheric pressure. Displacement is obtained from pump geometry and for rotary-vane pumps is twice the maximum crescent-shaped volume between the blades, inside the pump bore, multiplied by the revolutions per minute of the pump. The two times factor occurs since there are two pumping cycles per revolution. The volumetric efficiency of the rotary pump is typically 80 per cent and the pumping speed is the same for all permanent gases.

In operation the pumping speed will vary, depending on the inlet pressure. As the density of the gas in the system decreases, the mean free path increases,

* Fyrquel is made by Stauffer Chemical Company, New York, N.Y.
† Fomblin® is a registered trademark of Montedison S.p.A.
‡ Krytox® is a registered trademark of Du Pont Company.

Table 5.2 Typical applications and some physical properties of selected rotary pump fluids. (The blocked-in area denotes that the oil is suitable for that particular application)

	Hydrocarbon fluids.			Inert fluids
	Singly distilled medium viscosity	Doubly distilled	White oil	PFPE
Applications				
Mass spectrometers	■	■		
Electron microscopes	■	■		
Thin-film sputtering	■	■		
Surface studies	■	■		
UHV systems	■	■		
Leak detection	■	■		
TV tubes	■	■		
Power valves	■	■		
Distillation	■	■	■	
Space studies	■	■	■	
Furnaces	■	■		
EB welders	■	■		
Plasma etch			■	■
Impregnation	■	■	■	
Chemical pumping			■	■
Oxygen pumping				■
Radioactive				■
Packaging	■	■	■	
Mechanical booster				
Vapour booster				
Metallization	■	■	■	
Specifications				
Vapour pressure 20 °C (mbar)	$\leqslant 5 \times 10^{-5}$	2×10^{-5}	6×10^{-7}	2×10^{-6}
Molecular weight	490	400	430	1800 ± 100
Viscosity cSt at 20 °C	218	175	155	60
Pour point	$-9\,°C$	$-15\,°C$	$-12\,°C$	$-50\,°C$
Flash point	$234\,°C$	$213\,°C$	$243\,°C$	None
Fire point	—	$244\,°C$	$270\,°C$	None
Properties				
Energetic particle bombardment	Conducting polymers formed			No polymers formed (except with H_2 ions)
Thermal stability	Poor			Decomposes to gas alone
Oxidation resistance	Poor to fair			Excellent
Chemical resistance	Poor	Poor to good		Excellent
Radiation resistance	Fair	Fair		Good

Table 5.3 Relative oil price comparisons

Typical oil price comparison	Unit cost*
Hydrocarbon fluids (singly distilled)	1
Hydrocarbon fluids (doubly distilled)	1.3
White oils	5.7
Inert fluids (PFPE)	58.6

* Prices are based on 1 litre volume.

and the number of molecules occupying the suction volume of the pump goes down. In consequence, the rate at which gas is being removed from the system decreases. The pumps develop their full speed from atmospheric pressure down to about 0.1 mbar and the speed then decreases to zero at its ultimate pressure.

The limit on the ultimate pressure obtained is due to:

• The oil vapour pressure
• Back-leakage between the inlet and exhaust sides of the pump
• Air dissolved in the oil

Figure 5.7 shows typical speed curves (volume flow rates) for single-stage and two-stage pumps with and without gas ballast. Note that the two-stage pump produces an ultimate pressure two decades lower than the single-stage pump. Gas ballast increases the ultimate pressure of both pumps by more than a decade.

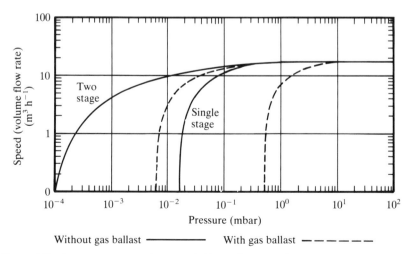

Figure 5.7 Rotary-vane speed curves (volume flow rates)

5.6 Rotary pump accessories

Inlet accessories

Catchpots
Catchpots are designed for fitting to the inlet of rotary pumps on applications where large quantities of vapours tend to condense in the pumping line and drain into the pump, or where froth or spray formed in the vacuum process can be carried into the pump.

Desiccant traps
Desiccant traps are used for the removal of small quantities of water at low pressures. The most common desiccant employed is phosphorus pentoxide (P_2O_5) which can absorb more than 20 per cent (by weight) of water. Rotary pumps that do not have a gas ballast facility can be protected with a desiccant trap against contamination of the pump oil by water vapour. The trap can be directly mounted to the pump inlet or fitted in the vacuum line at a convenient point.

Foreline traps
The nature of the gas flow taking place in the pipeline connected to the inlet of the rotary pump changes with pressure. At high pressure (atmosphere down to about 1 mbar) the high-density flow of gas prevents mechanical pump oil vapour from diffusing towards the system. However, as the pressure falls and the flow changes to 'transitional', the mean free path of the vapour molecules increases. Molecular collision is reduced and oil vapour may begin to back-migrate from the mechanical pump to the system, causing contamination.

Back-migration is the passage of vaporized molecules of the rotary pump oil from the inlet of the rotary pump towards other parts of the vacuum system.

Foreline traps have been developed for use on clean systems when it is important to prevent back-migration of rotary pump oil vapour into the vacuum system. The traps (see Figure 5.8) use activated alumina, which has exceptional trapping properties for oil vapour and has a long life before it becomes saturated and has to be renewed. During use the alumina becomes discoloured (yellow/brown) and when this discoloration can be seen extending about a third of the way through the alumina bed it should be changed. Fresh alumina should be heated to about 250 °C in air prior to placing in the foreline trap.

Activated alumina will absorb moisture; this may result in prolonged roughing times on successive pump-downs. Regular regeneration by baking

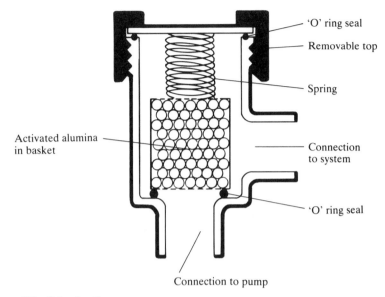

Figure 5.8 Inlet foreline trap

(or renewal) may therefore be necessary for rapid pump-down. Alternatively, a foreline trap changeover valve may be fitted in the roughing line, in parallel with the foreline trap. The valve, which is automatic in operation, permits the major portion of the atmospheric air to be pumped out of the system, bypassing the foreline trap and thus preventing early saturation of activated alumina by water vapour.

An alternative method is to use a liquid nitrogen cold trap in the pumping line to the rotary pump. This will condense and freeze out any back-migrating oil vapour. Obviously such traps must be operating (cold) whenever the system is pumping (say below 0.1 mbar) in order to perform. If an oil-saturated trap is allowed to become warm under vacuum it will itself become a source of oil vapour which can contaminate the system.

Another method is to utilize a dry nitrogen gas purge introduced into the pumping line. The inert gas flow thus increases the pressure in the pumping line and suppresses back-migration. When utilizing this arrangement in a diffusion pumped system, the backing pressure must be monitored to ensure that the critical backing pressure of the diffusion pump is not exceeded; otherwise the pumping action of the diffusion pump jets will be destroyed and gross contamination will occur.

Dust filters
Dust filters are required for applications where dust from a vacuum process may be carried over into the pump and cause rapid pump wear. The

impedance of a clean filter reduces the pumping speed by approximately 10 per cent at 1 mbar and 25 per cent at 10^{-1} mbar. Depending on the application, the filter element can normally be cleaned for reuse.

Chemical traps
Chemical traps provide protection against various aggressive vapours which may attack the pump or pump oil. In addition, they will prevent high molecular weight vapours, such as might arise in resin treatment plant, from reaching the pump, which in turn could cause lacquering or clogging.

The standard trapping material is activated charcoal, but other materials may be employed.

Exhaust accessories

Oil mist filters
Oil mist filters (see Figure 5.9) are designed to separate and trap the minute droplets of pump oil that may be carried out with the exhaust gases, expelled from the exhaust of rotary pumps. As such they reduce air pollution and act as effective silencers. On mobile installations, where venting into the working environment is necessary, the use of oil mist filters is recommended. Oil which collects in the filter can be returned to the pump. This is not advisable in cases where the oil may be contaminated. On large systems, returning oil can represent a considerable cost saving, especially where PFPE oils are being used.

The collected oil is normally returned to the pump through a pipe, which is connected from the drain of the filter to a low-pressure point on the pump, e.g. through a branch of the gas ballast connection or to the pump inlet.

Note that where completely odourless conditions are required, deodorizers can be recommended. These utilize activated charcoal to trap vapours associated with oil discharge and thus prevent oily smells.

External oil filtration systems
These filters are used to remove solid particles (typically down to 1 μm in size) and chemical contamination. Normally they feature an external oil pump which has sufficient pumping capacity for the oil in the rotary pump to be circulated through the filters every 5 to 10 minutes. The need to change filters is indicated on a pressure gauge supplied with the unit. Additionally these filters have a small oil capacity which allows more economical use of expensive oils such as Fomblin or Krytox.

Catchpots
As already discussed in Section 5.3, it is desirable, in some applications, for pumps to be provided with piped exhaust arrangements to carry gases and vapours to the outside of the building. There is always a risk that some

Outlet

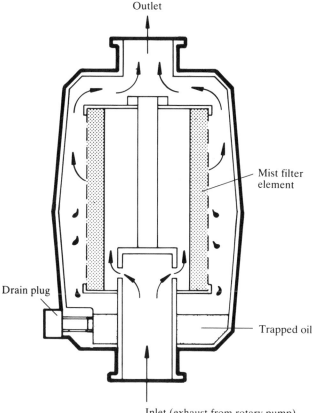

Mist filter
element

Drain plug

Trapped oil

Inlet (exhaust from rotary pump)

Figure 5.9 Exhaust oil-mist filter

vapour carried out with the effluent gas will cool and condense as it travels up
the exhaust line and that the liquid produced will drain back into the pump
causing serious contamination. For this reason it is good practice to use a
catchpot between the pump outlet and the exhaust line to prevent this
happening.

These catchpots are usually designed for fitting directly to the pump outlet,
or can be remotely mounted if required. A 'high level' sight glass is usually
provided to indicate the need for draining the catchpot. A drain plug is also
provided.

Vibration isolators

In some applications, especially when pumps are mounted in frames or
structures, antivibration mountings should be used to reduce noise and

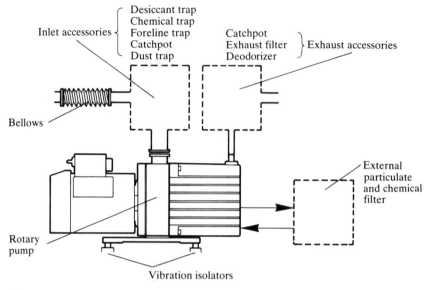

Figure 5.10 Rotary pump accessories

vibration transmission and to relieve stresses if mounting points are not level. (The use of flexible bellows or other flexible piping between the pump inlet/outlet connections is similarly recommended.)

The rotary pump accessories discussed are collectively illustrated in Figure 5.10.

5.7 Looking after rotary pumps

Some typical timescales are given for periodic maintenance; these refer to pumps used under average conditions. Pumps used in dirty or arduous conditions should be checked more frequently.

Daily or weekly	Check oil level
Half yearly	Change oil
	Inspect driveshaft for oil leaks
Annually	Inspect exhaust valve and gas ballast valve seals

Causes of poor performance

When properly used the mechanical rotary pump is extremely reliable and seldom breaks down. Therefore if a vacuum fault is indicated in the system, such as a poor vacuum, slow pump-down or no vacuum at all, the following should first be eliminated:

1. Possible misinterpretation of readings given by a vacuum gauge
2. Faulty vacuum gauge
3. System leak
4. System contamination
5. Contaminated rotary pump oil

Only then should the pump be investigated and if necessary dismantled or sent for repair. Item 5 is by far the major maintenance problem. A visual check of the oil may be useful; it may appear 'milky'. A change of oil may therefore resolve the problem.

Although mentioned earlier, it is again worth stressing that suitable precautions must be taken when working on contaminated pumps. These will depend on the nature of the materials involved, but particular attention must be paid to the danger of inhaling vapours or skin contact with the oil or sludge/debris from process materials.

During an oil change never use a solvent to 'flush' out the remaining oil dregs; where possible use a low-vapour-pressure flushing oil recommended by the manufacturer. Solvents could possibly attack some of the pump seals; remains of the solvent can dilute the fresh oil and its high vapour pressure will make the subsequent attainment of vacuum difficult/prolonged.

If a change of oil does not resolve the situation then a mechanical fault is probable. Faults may be any of the following:

1. *Faulty valves*, e.g. gas ballast, exhaust, oil circuit, inlet isolation, etc., as and where applicable.
2. *Shaft seals and 'O' rings*, if used in pump construction, may fail through general wear and tear.
3. *Wear*, e.g. blades, rotor, stator, springs. Wear and breakage defects are very unlikely, except after very long use.
4. *Breakage*, e.g. springs.
5. *Belt slippage* for pumps with belts.
6. *Incorrect assembly* by maintenance personnel not following the correct procedure.

The performance of a rotary pump depends largely on good surface finish and effective sealing within the bore of the pump. Production of a seal that will withstand the pressure differences encountered within the pump requires that the gaps between the blades of the rotor and the bore of the stator as well as the gap between the rotor and stator itself must be very small. Blade tips are normally self-adjusting for wear, being spring loaded against the stator bore. The blade-to-rotor slot gap is typically 0.04 to 0.1 mm (0.0016 to 0.004 in). If blades are replaced and are too tight they may 'stick' in position. Too big a gap as a result of wear or replacement with the wrong blades will allow excess leakage. Similarly, damaged blade tips can lead to leakage between the exhaust and inlet side of the pump. Damage such as 'scoring' to the rotor of the pump and especially to the stator in the region where it is close to the rotor can be particularly detrimental to performance.

In the majority of rotary pumps the stator has an arc machined in the bore surface at the point where the rotor fits closely (see Figure 5.11). Often known as the 'duo-seal', the arc has a radius equal to the radius of the rotor, with the rotor so positioned relative to it to give a comparatively long clearance between them. The gap between the rotor and the duo-seal is of the order of 0.025 mm (0.001 in). Any damage to the stator surface would have to extend across the whole duo-seal surface to be significant. This design has obvious advantages over tangential (stator/rotor) line contact.

In some pumps, particularly older types, the duo-seal clearance needed to be set manually during manufacture and service. In more modern pumps the pump is assembled on dowel pins that locate the component parts correctly, and manual setting is therefore not necessary and subsequent servicing of the pump is easier.

Dismantling and assembly of the older manually set pumps needs to be

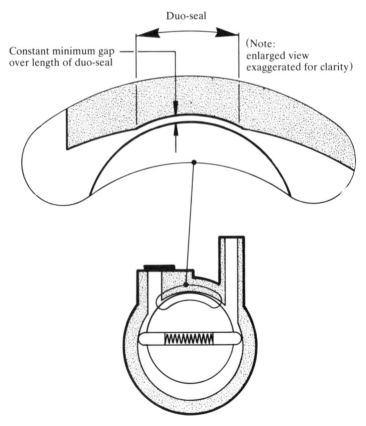

Figure 5.11   Rotary pump duo-seal

carried out with care following the manufacturer's recommended procedures. If the pump is set with too small a clearance there is a possibility of pump seizure due to expansion of the parts at running temperature. Too large a clearance leads to excessive back-leakage of gas from the exhaust side of the pump back into the inlet region.

Figure 5.12 gives a fault-finding flow diagram which can be used generally for oil-sealed rotary-vane pumps incorporating gas ballast facilities.

On completion of pump reassembly, it is advisable to test the pump to check its ultimate pressure performance before putting it back into use. The correct procedure for doing this is given in the figure. The times quoted may seem excessive if the pump is required urgently, but they are necessary if true ultimate pressure measurements are to be obtained. However, the time can be reduced provided it is realized that the true performance is not being measured.

For quick testing the pumps should be run isolated for $1\frac{1}{2}$ to $2\frac{1}{2}$ hours on full gas ballast. The gas ballast valve should then be closed and the pump connected to a McLeod gauge. Readings can be made after a further $\frac{1}{4}$ hour. Compare the results obtained with those quoted by the pump manufacturer.

A Pirani-type gauge can be used for this test but the results obtained will be considerably higher, unless the gauge is trapped (i.e. with liquid nitrogen) to prevent oil vapours from entering it. The Pirani gauge measures total pressure and will thus include a partial pressure component for any oil vapour that may back-migrate from the rotary pump. Because of the compression effect on the gas sample in the McLeod, it will only measure the partial pressure of the uncondensed gases in the sample. The partial pressure of oil vapour will be excluded from the measurement.

To obtain readings of pump inlet pressure the gauge itself must be clean, as must be the pipework connecting it to the pump. The pipework must be free of leaks and should be as short a length and as large a bore as possible. Vacuum hose may be used to join pipe lengths but must be kept to minimal length.

5.8 Rotary pumps for semiconductor applications

As mentioned briefly in Section 1.5, one of the most demanding application areas for pumping systems incorporating rotary-vane pumps has been in the semiconductor industry.

Typically, the prime attack is on pump oils. Mineral oils can be attacked to produce thick, very viscous, or even solid products (see Figure 5.13). These solids may obstruct oilways which provide lubrication paths into the pumping mechanism and additionally cause wear in bearings and journals. Oil filtration is generally essential for reliable operation. The use of an external oil recirculation system with activated alumina cartridges removes both acid and particulate matter. Such systems incorporate oil pressure

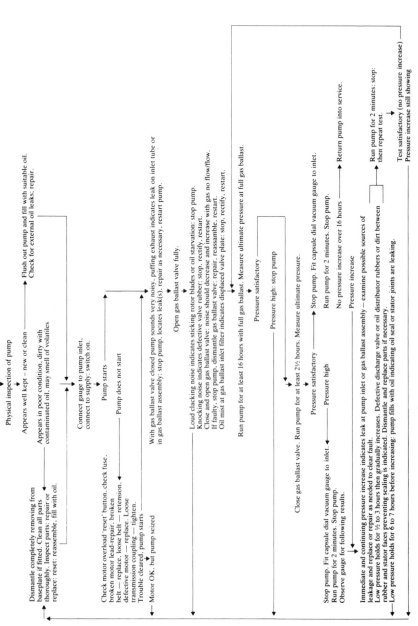

Figure 5.12 Oil-sealed rotary vane pump fault-finding flow diagram

Figure 5.13 A rotary pump used in a low-pressure chemical vapour deposition application. Clearly, considerable particulate matter has formed inside during processing. Oil filtration is thus essential.

monitoring and audible alarm facilities to give advance warning of maintenance requirements. Synthetic oils, such as the perfluoropolyethers (PFPE), resist chemical attack, but if acidic fluids are allowed to accumulate within the pump and the pump is switched off, their concentration on top of the oil can produce fierce local attack of the metal surfaces.

Vapours such as carbon tetrachloride or carbon tetrafluoride are solvents and if they condense in the pump will dilute mineral oils and impair their lubricating properties. Again PFPE oils are recommended. Additionally, in such cases it is advantageous to operate with a gas ballast of dry nitrogen (air containing water vapour is likely to promote acidic attack) so that the solvents and acids are prevented from condensing in the pump and pass through to the exhaust line where they may be treated appropriately.

A different type of problem that must also be considered is the pumping of oxygen in high concentrations. Explosions have sometimes occurred when rotary pumps lubricated with mineral oil have been used in applications, such as photoresist stripping, where oxygen is present. Atomized oil mist, hot oil vapour, friction, compression, elevated temperatures and possible static electricity, combined with oxygen, result in a potentially hazardous environment. Most of these conditions will be found within a vacuum pump.

The risk of explosion occurring can be minimized by introducing an inert gas, such as nitrogen, into the pump in the form of gas ballast, and also into the oil box reservoir to act as a purge, and additionally reducing the concentration of oxygen to less than 25 per cent. The main safety problem with this practice is to maintain constant gas flow conditions. For instance, as the supply bottle empties the inert gas supply may be disrupted or cease. Also, there is the possibility of a failure in the oxygen supply line which could result in high flow conditions, a rise from low pressure to atmosphere or even cylinder pressures. Current practice includes fitting monitors to the oxygen and nitrogen lines which will shut down the vacuum pump in the event of such changes taking place. The potential problems do not occur only in the pump itself but may occur with mineral oil deposition in exhaust lines, causing a serious fire hazard if reaction occurs with certain exhaust gases.

The majority of semiconductor process equipment manufacturers in the world now specify PFPE oils for plasma etching and photoresist stripping where oxygen is present. Also, the majority of end users in the semiconductor industry use these oils to maximize safety and minimize maintenance.

The sources of contamination (problems) that rotary pumps can be subjected to when used in semiconductor processing plant and their effects are summarized in Table 5.4.

Figure 5.14 shows a mechanical vacuum pumping system which, although designed primarily for the highly aggressive plasma etch process, can be used in many other corrosive applications. It incorporates many of the features discussed and in particular has the following design innovations:

1. A positive pressure lubrication system, powered by an internal shaft-mounted circulation pump, ensures that all surfaces that may come into contact with aggressive gases are protected by a layer of perfluoropolyether fluid.

   It is known that the vast majority of pump failures are caused by blockage of the lubrication circuit. A continuous supply of clean lubricant is therefore necessary to ensure that oilways do not become blocked by particulates.

Table 5.4 Problems for rotary pumps used in semiconductor manufacturing

Sources of contamination	Effects
Particulates	Abrasion of surfaces, e.g. bearing surfaces
	Blocking of lubrication channels
Acid and corrosive materials	Attack on materials of pump
	Reaction with mineral oils
	Concentration on top of inert oils
	Toxic hazard
Solvents	Dilution or polymerization of oil
	Attack of elastomers
	Reduction of vacuum performance
Oxygen and hazardous gases	Explosive hazard with pump fluids
	Toxic hazard in certain cases

Figure 5.14 Vacuum pumping system for plasma etch process (the mechanical booster pump is an option)

In this system the return line from an external filtration system feeds filtered, deacidified lubricant directly to the shaft-mounted circulation pump. As a further safeguard an additional filtered inlet is situated in the oil-box so that the pump can continue working if the external filter is stopped for short periods for routine changing of filter elements. Stainless steel is used for all internal lubricant supply lines.

2. The external filtration system has a motor-driven oil filter which features a sealed pump with magnetic drive. The connecting hoses are armoured Teflon with stainless steel quick-disconnect unions for faster filter replacement. The filter elements for aluminium etch applications are activated alumina-impregnated rolls contained in Teflon-coated cannisters, which scavenge the acid from the perfluoropolyether fluid. A pressure gauge fitted to the system helps to indicate when cartridge replacement is necessary. PFPE fluid can be recovered from the used elements by means of a filter press.

3. The rotary pump incorporates a specially designed shaft seal arrangement shown in Figure 5.15. The most probable failure of a standard pump is

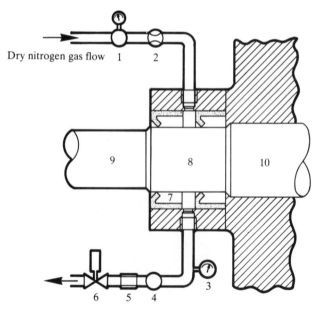

Dry nitrogen gas flow

KEY

1	Pressure regulator	6	Magnetic valve
2	Flowmeter	7	PTFE shaft seals
3	Pressure gauge	8	Replaceable hardened shaft sleeve
4	Fixed restrictor	9	Shaft to motor dive
5	Clear Teflon hose	10	Pump rotor

Figure 5.15 Double rotary seal arrangement employed in the rotary pump

likely to be due to damage to the shaft seal normally caused by residual particulates. PFPE oils, having very low surface tension, will seek out the most minor shaft seal deficiencies and will thus leak through a shaft seal more readily than the standard hydrocarbon oils. A solution offered in this system is to use a double shaft seal arrangement. The inner seal is a positive action PTFE seal while the outer is a normal PTFE lip seal. Note the replaceable hardened shaft sleeve—item 8. Dry nitrogen is bled into the space between the two shaft seals and then via a restriction to the gas ballast inlet of the pump.

Normally the pressure in this space is about 0.2 bar (above atmospheric pressure) and even with a damaged seal, with this positive pressure, oil leakage is unlikely. If, however, any oil does pass the first seal it is then entrained and returned by the gas ballast flow into the gas ballast inlet. The return line is translucent so that any oil returned is visible and preventive action can be taken. A solenoid valve in the inlet to the gas ballast assembly ensures that this circuit is closed when the pump stops.

4. Other features of the system include dual regulators and flowmeters for nitrogen gas ballast flow and oil-box purge gas flow together with oil pressure indication.

6

Oil-free mechanical primary pumps

6.1 Mechanical booster pumps (Roots pump)

Figure 6.1 shows the cross-section through a typical mechanical booster pump where 'figure-of-eight' rotors (often referred to as impellers) are synchronized by external gears. The rotors rotate in opposite directions inside a stator and do not touch either each other or the stator walls. The clearance is generally 0.1 to 0.5 mm (0.004 to 0.020 in) when cold. While oil is used in the pump, the pumping compartment is oil free. The working principle of the mechanical booster pump is also illustrated in Figure 6.1. Gas is trapped between each rotor and the stator wall and transferred from the inlet to the exhaust side of the pump. The pump operates at rotational speeds typically between 1400 and 4000 rev min^{-1}.

As a consequence of the clearances between pumping components, back-leakage of gas occurs at a rate governed by the pressure difference between the input and output (compression ratio) and the type of gas being pumped. Exhausting to atmosphere gives a ratio of about 3.5:1, which allows a fine side pressure of the order of 300 mbar to be achieved. In order to obtain lower pressures the booster is normally used with the exhaust side connected to the inlet of a backing pump, such as a rotary-vane pump. A compression ratio of 50:1 is then typically obtained for backing pressures of 0.05 mbar.

Mechanical boosters are normally used in combination with rotary pumps.

Typically, the mechanical booster is employed for pumping vacuum melting furnaces, in impregnation plant for electrical equipment and in low-density wind tunnels. A range of sizes are available with pumping speeds typically between 150 and 50 000 m^3 h^{-1} operating mainly in the pressure range of 10 to 10^{-3} mbar. The peak pumping speed is developed in the pressure range of 1 to 10^{-2} mbar, the speed at the lower end of the pressure range depending on the performance of backing pump used. Figure 6.2 shows a selection of speed curves for various rotary pump and mechanical booster combinations.

Mechanical booster/rotary pump combinations can remove large volumes of gas in the medium-vacuum range.

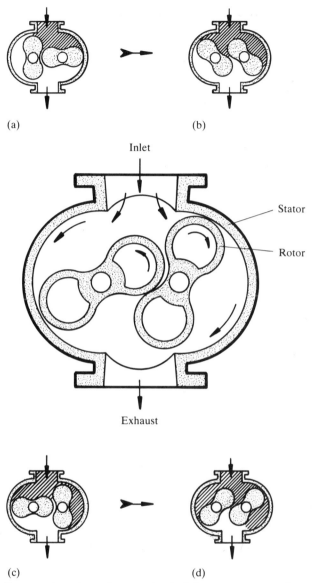

(a)   (b)

Inlet

Stator

Rotor

Exhaust

(c)   (d)

Figure 6.1   Cross-section through a mechanical booster pump and stages in its operation

Figure 6.3 shows a cutaway view of a booster pump. Typically the stators and rotors of small pumps are made of cast iron; larger rotors can be fabricated from steel and are usually dynamically balanced. Gears and bearings are in housings external to the main stator and the shafts pass through shaft seals in the ends of the stators. Bearings are normally ball or

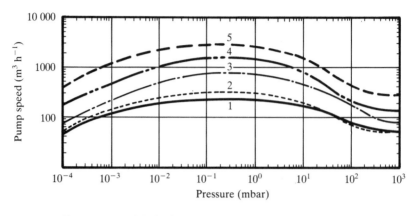

	Two-stage rotary pump	Mechanical booster
1	40	250
2	40	500
3	80	1200
4	175	2600
5	275	4200

Numbers are pump speed in $m^3 h^{-1}$

Figure 6.2 Speed curve for various rotary pump and mechanical booster pump combinations

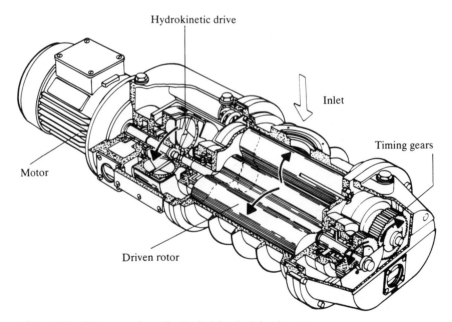

Figure 6.3 Cut-away view of a hydrokinetic drive booster

roller types and are lubricated by oil or grease. Commonly a pumping duct is provided to link the housings to the discharge side of the pump; thus during operation there is less tendency for oil or grease to be forced through the shaft seals into the pumping compartment due to the pressure differential between them.

In Figure 6.3 the external timing gears can be seen, housed at one and of the pump. At the opposite end of the pump the driveshaft extends through a further shaft seal in the housing to protrude outside the pump for connection to the electric motor. In some pumps the gears are housed at the motor end of the pump.

6.2 Hydrokinetic drive boosters

The pressure differential across the booster pump must be carefully controlled for two reasons:

1. Severe heating (particularly of rotors) can occur with large gas loads at high compression ratios. The resulting temperature rise and expansion can lead to closure of working clearances and consequent seizure. This problem may be reduced by cooling the rotors with oil passing through their shafts or more conveniently by providing a water-cooled heat-exchanger in the gas discharge area close up to the rotors.
2. The torque input to the mechanical booster is dependent upon the pressure difference across it. At full rotational speed this becomes unacceptably high above about 10 mbar inlet pressure, depending on the displacement ratio of the mechanical booster and backing pump. If the torque is allowed to exceed the rated output of the motor, then high current will flow in the windings which will cause the motor starter to trip.

There are several ways of overcoming this difficulty:

1. By pressure switching such that the mechanical booster pump is only switched on below 10 mbar. However, if the system is roughed out through the booster, its rotors act as an impedance to the backing pump, reducing its effective pumping speed. This can be overcome by providing a bypass line with a valve that is open only when the mechanical booster is not running. This arrangement is obviously more costly.
2. By providing a pressure relief bypass valve such that the maximum pressure differential that the motor can drive is not exceeded. Back-flow through the bypass from the outlet to inlet limits the pressure difference across the pump during the early part of the pump-down from atmospheric pressure. The bypass valve closes automatically after the backing pump has brought the pressure difference below a safe level. However, recirculation of gases that have already been pumped once is a disadvantage. Additionally there is a possibility that the valve may stick under dirty pumping conditions.

3. By providing hydrokinetic drive between the motor and pump so arranged that full load torque is not exceeded. The pump automatically slows down at high pressures, thereby preventing the pump from overheating or the motor from overloading.

The power transmission unit of the hydrokinetic drive consists of a driving coupling (see Figure 6.4) mounted on the motor shaft and a driven coupling on the rotor shaft. The two halves of the coupling are not connected to each other and do not touch each other, the drive being transferred by oil circulation alone. In operation, the oil is thrown up from the reservoir by the spinning–driving coupling. It provides spray lubrication and cooling for bearings, etc., and gives up heat to the finned oil cooler. Much of the oil spray is caught in a trough and fed by gravity into the space between the driving and driven couplings of the transmission unit. It is centrifugally circulated from one half of the coupling to the other, transferring angular momentum and torque before spilling out and back to the reservoir.

Such a coupling has a characteristic that it never transmits more than a selected maximum torque. This torque is selected so as never to overload the drive motor. High torque is transmitted whenever the driving and driven transmission couplings are running at markedly different rotational speeds; transmitted torque falls to a low value when the speed of the driven coupling approaches that of the driving coupling.

When switching on, even a large pump, the motor runs up to speed easily without any special starting arrangements. The pump can be switched on at the same time as its backing pump and immediately begins to assist the pumping process from the very beginning of pump-down. Pump-down time to a particular pressure can be greatly reduced and there is no need for a pressure switch, bypass or any external arrangement or control normally

Oil cooler

Figure 6.4 Hydrokinetic fluid coupling built into the mechanical booster. The motor and driving half of the coupling are on the left, the pump and driven half of the coupling are on the right

associated with mechanical booster systems. The pump is well protected against operator error or accident. Even if the backing valve is left closed or the backing pump switched off, the booster pump will merely rotate slowly without motor overload. A sudden inrush of atmospheric air into the inlet merely causes the rotors to slow down while the motor continues to turn at normal rotational speed. If solid debris enters the pump and causes jamming or seizure of the rotors, these merely remain still and produce no overload of the motor.

Figure 6.5 shows pump-down characteristics of a particular booster pump/backing pump combination employing hydrokinetic drive, compared with the characteristics for the same size of pump employing pressure switching, where the mechanical booster only begins to contribute at a late stage during the pump-down process. The value of the booster contribution during the whole of the pump-down period is obvious.

A hydrokinetic drive booster/rotary pump combination can give a significantly reduced system pump-down time compared with the rotary pump alone.

Routine maintenance of mechanical boosters includes the regular inspection and maintenance of all the lubricants (bearings and gear oil housings and shaft seal oil reservoirs, etc.). Rapid disappearance of oil from the oil seal

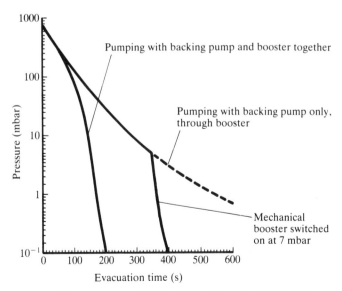

Figure 6.5   Pump-down curves of a 2.8 m³ chamber obtained with a 220 m³ h⁻¹ backing pump and 2500 m³ h⁻¹ mechanical booster, the latter driven by a fluid coupling

reservoir would indicate that the shaft seal may need changing. Oil levels are normally checked when the pump is not running. Mechanical boosters are normally reasonably quiet-running machines, but small changes in synchronization or in gears or bearings over a long period of time may cause noisy operation.

6.3 Magnetic drive boosters

In the uranium enrichment industry, the separation of uranium isotopes is achieved using a large number of gas centrifuges connected together in cascade. A gaseous compound of uranium, uranium hexafluoride (UF_6), is used. Prior to start-up, each centrifuge is evacuated and sealed off, and during processing mechanical boosters are used to remove and compress the gas prior to storage. The cascade operates at a pressure of a few millibars.

UF_6 is a highly reactive material. The major problems of vacuum pumping UF_6 are associated with the presence of hydrocarbons and possible leakage of gas. UF_6 reacts with any material of a hydrocarbon nature and may produce a byproduct that is a hard solid. If this solid is produced in a fine particulate form it can be highly abrasive for the pump. Water vapour in an air leak, leaking into the vacuum system, would be a problem, since the presence of hydrogen would give rise to similar reactions to those with hydrocarbons. For these reasons mineral oil as a pump lubricant is precluded and perfluoropolyether (PFPE) fluids and greases are used. Seals containing hydrocarbons cannot be used. Others containing chlorine or silicone would also be attacked by UF_6. However, a fluoroelastomer seal (e.g. Viton) or polytetrafluoroethylene (PTFE) seal is suitable and not attacked.

With normal mechanical boosters pumping UF_6 gas, there would be a danger of gas leakage through the dynamic shaft seal.

There are a number of possible solutions to be problem:

1. Seal the complete motor into the vacuum. Although useful in some vacuum applications, in this case having the windings in such an aggressive environment would present insurmountable problems.
2. A commonly used solution is to use a 'canned' motor, where a very thin (0.5 mm) non-magnetic metal 'can' or tube surrounds the electric motor rotor, the stator and windings being outside the 'can'. This does away with the need for a shaft seal. Drive is achieved by magnetic induction between the stator and rotor through the 'can'. However, there are problems. To obtain reasonable efficiency from the motor the total gap between rotor and stator must be kept small. The small gap between rotor and 'can' could be bridged with abrasive particles, causing seizure or 'can' rupture.
3. An alternative solution to this problem is to use a leak-tight magnetic coupling between the motor and driven rotor. The magnetic coupling,

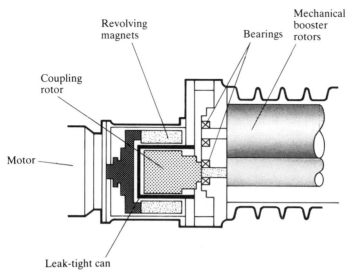

Figure 6.6   Leak-tight magnetic drive arrangement for mechanical boosters

shown in Figure 6.6, consists of a rotor similar to an induction motor rotor, surrounded by a substantial non-magnetic 'can' (the gap between rotor and can is about 1 mm). A ring of permanent magnets is rotated around the outside of the 'can', thus producing a rotating magnetic field. Drive is achieved by magnetic induction between the rotating magnetic field and the induction rotor through the 'can' wall. Magnetic coupling has many advantages. The thick 'can' offers much greater security. The gap between rotor and 'can' is too large to be bridged by small particles. The rotating magnets act as a cooling fan giving powerful forced cooling. Any motor, even an air or hydraulic motor, may be used.

6.4 Direct drive boosters

An expanding field for the mechanical booster is in its application in fast axial gas flow CO_2 power lasers for welding and metal cutting purposes. In this application the booster acts as a gas circulator. The gas required for operation is admitted to the laser in the pressure range of 50 to 200 mbar. The consumed gas is automatically replaced by 'fresh' gas. The need for a constant flow rate requires direct drive mechanical boosters driven at constant rotational speeds. These are especially adapted to run with their inlet and exhaust ports in a horizontal plane for compactness. Because the pumps are required to run at a constant, relatively high, pressure, they have larger motors and some pump clearances are increased.

6.5 Variable-speed drive boosters

Control of pressure at a predetermined level can be a requirement in research and industrial vacuum systems, particularly in the semiconductor industry in processes such as low-pressure chemical vapour deposition and metal etch.

A variable-speed drive mechanical booster combination (with hydrokinetic drive) offers adjustment of pumping speed in response to changes in operating pressure via a drive motor frequency controller, used in conjunction with a pressure gauge and pressure set-point instrument. By varying its pumping speed the mechanical booster can be made to maintain constant system pressure, even under conditions where gas load may vary substantially.

Variable pumping speed is achieved by using a motor frequency controller to adjust rotational speed. The frequency controller, in turn, is adjusted by a pressure set-point controller continuously receiving system pressure data from a pressure gauge.

6.6 Drypump systems

It was shown in Section 5.8 that, in the semiconductor industry, rotary-vane pumps are subjected to the most demanding operating conditions. Pumps and systems must handle process gases plus reaction by-products, which may be gaseous, liquid or solid, and may be:

- Corrosive
- Toxic
- Reactive
- Condensable
- Pyrophoric
- Flammable
- Abrasive
- Sludge forming

In particular, abrasives cause wear of surfaces within pumps (e.g. bearing surfaces), and may cause blocking of lubrication channels. Acids and corrosive materials may react with mineral oils; they may attack some of the materials of construction of the pump; they may concentrate on top of inert fluids, causing fierce local attack of materials; and if retained in the pump they may present a health hazard to service personnel. Solvents may dilute or polymerize the oil; they may attack elastomers within the pump and have an adverse effect on the vacuum performance. Oxygen and other hazardous gases may present an explosive hazard with pump fluids or a toxic hazard in certain cases. Most of the problems are thus associated with the oil in the pump.

The key considerations when selecting a rotary pump for semiconductor applications are:

- Safety
- Clean vacuum
- Low running costs
- Reliability
- Good vacuum performance
- Minimum maintenance

Using oil-sealed rotary pumps it is difficult to optimize each of the above requirements, so a compromise must be accepted. This frequently results in high running and maintenance costs. It was these problems that have promoted the development of pumps capable of operating from atmospheric pressure down to around 10^{-2} mbar, which work without oil in the swept volume, i.e. drypumps.

Potential dry pumping mechanisms for consideration are:

- Carbon vane blades
- Screw rotors
- Roots stages
- Claw rotors

Carbon vane pumps cannot offer an adequate vacuum performance for the majority of semiconductor applications and are known to be sensitive to particulates.

An oil-free screw rotor-type pump needs to be multiple stage and to rotate at very high speeds (10 000 rpm +) in order to be effective and produce a useful performance. This leads to problems with vacuum leak-tight sealing of shafts, as well as a relatively complicated gas path, with the associated reliability problems.

The Roots principle is well known and is especially effective when delivering against low-pressure differentials. However, to deliver to atmosphere, a complex design involving several stages and interstage cooling with gas recirculation is required. Apart from a high power consumption, there is a significant risk of condensation and particulate accumulation between stages.

The claw mechanism can deliver efficiently against high-pressure differentials to atmosphere, thus removing the problems of interstage cooling or gas recirculation. A pump consisting only of claw-type stages would be suitable down to pressures around 10^{-2} mbar. However, the Roots mechanism is more efficient in the lower pressure region.

A development by one vacuum manufacturer is to combine Roots and claw mechanisms on intermeshing rotor pairs mounted on common shafts and held in correct phase relation without contact by timing gears (see Figure 6.7).

A four-stage version is capable of an ultimate pressure of 10^{-2} mbar without gas ballast and 3×10^{-2} mbar with gas ballast. High gas ballast flow rates are possible since no carryover of oil mist will occur; the resulting high

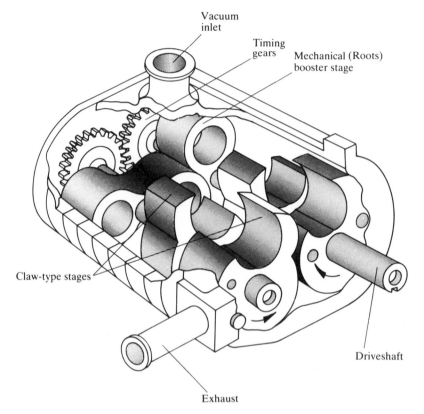

Figure 6.7 An oil-free mechanical pump capable of operating from atmospheric pressure down to 10^{-1} mbar

gas flow rate helps to prevent particulate formation and solids pass to the exhaust where they can be handled more simply. The risk of explosion or corrosion is reduced. Water cooling allows precise control of the pump operating temperature. Pump performance improves as the pump warms up (15 to 30 min).

Depending on the requirement of the application, the pump can be fitted with an electronic control system to provide safety features (see Figure 6.8). Safety features include an automatic system shut-down, which stops the pump when operating temperatures exceed preset limits, preventing pump damage or thermal seizure. Another safety feature is an audible alarm which warns the operator that exhaust passages are constricted and require immediate maintenance. If the exhaust restrictions are not corrected and exhaust back-pressure increases, the system automatically shuts down and cannot be restarted until the necessary maintenance is performed. The controller also permits automatic system start-up and shut-down for

Figure 6.8   An oil-free mechanical pumping system with soundproof enclosure removed to reveal a four-stage dry pump at the bottom with an additional mechanical booster mounted on top. Such a system, known as a Drystar system (manufactured by Edwards High Vacuum International), is capable of reaching pressures into the low 10^{-3} mbar range

untended operation. A mechanical booster is shown situated above the drypump in the illustration. Such an arrangement will extend the ultimate pressure to 10^{-3} mbar.

The whole system is normally enclosed in a soundproof enclosure—low noise level being one of the environmental requirements of a modern pump. A fan located in the housing helps to remove heat.

Many international semiconductor manufacturers in the USA, Japan and Europe have demonstrated the success of this particular oil-free pumping system. Field experience confirms that oil-free pumping systems offer a number of advantages. Reliability is improved and maintenance and running

Table 6.1 Key comparisons with regard to servicing oil-sealed rotary pumps and drypumps used in the semiconductor industry

	Traditional oil-sealed rotary pumps	*Drypumps*
Servicing	Preventative maintenance requires frequent exposure of service operator to risks of: Contaminated oily parts Trapped hazardous gases Hazardous sludge	Significantly less maintenance, hence less exposure to risks
Disposal of waste	Contaminated oil disposal/regeneration Contaminated oil filter elements Contaminated exhaust mist filter elements	No oil in swept volume No oil filter No mist filter

costs are significantly reduced (payback times at 10 months to 1 year have been quoted).

A summary of key advantages of drypumps, with regard to servicing and waste disposal, compared with conventional oil-sealed rotary pumps is shown in Table 6.1.

Drypump technology introduces a new concept in process pumping, providing users with the opportunity to redesign vacuum systems and achieve significant improvements in safety, uptime, reliability, cleanliness and running costs. In this respect drypumps should be ideal pumps for some uses in the chemical process industry (for fatty acid distillation). Additionally drypumps have been used in freeze drying applications because of their cleanliness.

7

Diffusion pumps and accessories, integrated vapour pumping groups and vapour boosters

7.1 Introduction

The most widely used pump for obtaining high vacuum is the diffusion pump. Invented in 1913, the diffusion pump has enjoyed widespread and increasing use both in industrial and research applications. Although its use has been mainly associated within the high vacuum range, diffusion pumps today can produce pressures approaching 10^{-10} mbar when properly used with modern fluids and accessories.

In general, the features that make the diffusion pump attractive for high and ultra-high vacuum use are its high pumping speed for all gases and low cost per unit pumping speed when compared with any other type of pump used in the same vacuum range.

7.2 Mode of operation

The diffusion pump is incapable of exhausting directly to the atmosphere and requires a backing pump (generally a rotary pump) to be operated continuously in series with it. A schematic diagram of a diffusion pump is shown in Figure 7.1. For simplicity only one jet stage is shown; in practice actual pumps may have up to five jet stages. The outside of the cylindrical body of the pump shown is water cooled, although air-cooled versions incorporating a fan blowing over fixed fins are available, usually for portable use. An electrical heater is clamped against the underside of the base of the pump and a working fluid is placed inside the base. This fluid is normally a low vapour pressure oil although some versions of diffusion pump use mercury. Inside the body of the pump is a hollow jet assembly, often referred to as the chimney.

Operation is as follows: with the inlet of the diffusion pump normally closed by a valve, the rotary pump is switched on and left running continuously, a pressure of at least 0.1 mbar being required on the exhaust side of the diffusion pump. The water supply or air cooling is turned on. The oil can now be heated and within perhaps 15 to 20 minutes begins to

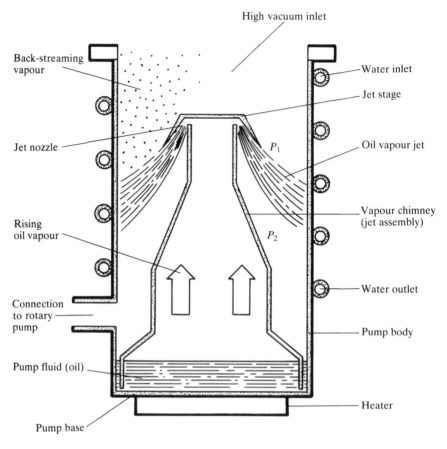

High vacuum inlet

Back-streaming vapour

Water inlet

Jet stage

Jet nozzle

P_1

Oil vapour jet

Vapour chimney (jet assembly)

Rising oil vapour

P_2

Connection to rotary pump

Water outlet

Pump body

Pump fluid (oil)

Heater

Pump base

(Note: Pressure P_1 is less than P_2. Water flow is down the pump.)

Figure 7.1 Basic elements of a diffusion pump showing only one jet stage

boil. The hot oil vapour formed fills and escapes from the jet nozzles, in the direction shown in Figure 7.1. The vapour expands in passing from a region of comparatively high pressure (the boiler pressure inside the jet assembly is of the order of 3 mbar) to one of low pressure. An annular vapour jet is created which is moving at a velocity that is supersonic and impinges on the inside of the water-cooled outer wall. Here the vapour condenses and the oil flows down the wall and is returned to the boiler for recycling. Gas molecules diffusing into the pump mouth are likely to collide with the much heavier oil vapour molecules and will most probably receive a velocity component directed towards the exhaust side of the pump, where they will be subsequently removed from the diffusion pump by the rotary pump. A pressure difference is thus established across the continuous flowing vapour jet.

The diffusion pump itself cannot be exposed to the atmosphere while the fluid is at working temperature, because of the problems involved with fluid degradation and system contamination, but suitably valved and operated, it can be cycled almost indefinitely.

7.3 Back-streaming

In any vacuum system where a chamber or vessel is connected directly to a diffusion pump, backed by a rotary pump, the pressure in the chamber will eventually reach a point of equilibrium at which continued operation of the pumps will not further reduce the pressure of the system. The pressure in a leak-tight evacuated vessel is limited by the outgassing load from the chamber. It is also limited by the presence of oil vapour derived from the diffusion pump. The source of this oil vapour can be grouped into two categories: back-streaming and back-migration:

1. Back-streaming is the direct flight of vapour molecules from the pump nozzle (particularly that of the top stage) towards the mouth of the pump.
2. Back-migration is the transfer of vapour to the high vacuum side by the reevaporation of the fluid molecules which cling to surfaces within the pump, particularly the pump body adjacent to the top stage. Back-migration is thus temperature dependent and can be reduced by arranging the cold water inlet feed to supply this region first.

Back-streaming is more difficult to suppress; it may be reduced by careful nozzle design, but usually reduction by this procedure is at the sacrifice of the pumping speed. The source of back-streaming associated with the top stage can be subdivided into two main components:

1. *Wet running lip*. Droplets of liquid pump fluid can sometimes collect around the bottom edge of the top jet cap, providing sources for random evaporation. This condition is referred to as 'wet running'.
   Consider the right-hand side of the top stage jet cap shown in Figure 7.2a. Wet running is shown on the underside with a drop of liquid on its lip. This condition can occur either by direct splashing from the boiler or by condensation of the vapour inside the top cap. Evaporation of droplets on the cap lip can take place, and much of the vapour travels into the system out of the pump mouth.
2. *Turbulent boundary layer effect*. This is shown on the left-hand side of the cap. The vapour jet emerging from the nozzle has been affected by frictional drag along the nozzle walls. This leads to a boundary jet layer consisting of slow-moving molecules, some of which after intercollision go towards the chamber.

Back-streaming sources associated with the top jet can be significantly reduced by using a guard ring or cold cap, shown in Figure 7.2b. The cap is

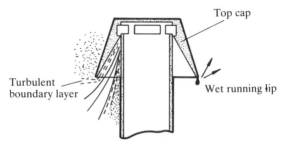

(a) Sources associated with the top stage

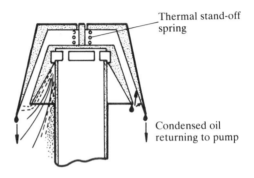

(b) Cold cap covering the top stage

Figure 7.2 Back-streaming sources from the top stage and their reduction

either water cooled or conduction cooled from the top flange and its lower edge is just lower than that of the top stage it surrounds. The cooled surface of the cold cap intercepts vapour back-streaming from the top jet and causes it to condense. The condensed oil formed on the cap eventually drops off and returns to the boiler. Using this device can give a reduction in back-streaming of up to 100 times that obtained without a cap. (Note: In further discussion, the use of the term back-streaming will be taken to include back-migration.)

7.4 Fractionation in oil diffusion pumps

Diffusion pump oils do not have a single molecule size; they have a range of molecular weights typically between 300 and 600 a.m.u., compared to the mercury molecule of atomic weight 201. Each component may have a different boiling point and vapour pressure. The ultimate pressure will be governed by the highest vapour pressure component present in the region of the pump inlet. An improvement in ultimate pressure can be obtained if the vapours in this region (which generally come from the top jet) contain a minimum proportion of high vapour pressure fractions.

There are two main ways in which this reduction can be achieved: one is to remove light fractions through purification of the oil in the pump; the other is to ensure that high vapour pressure components are fed to the lower stages and not the top stages of the pump. This can be achieved through fractionation of the oil in the pump.

Fluid purification (the ejector stage)

Every pump has to a certain degree a self-purifying tendency. The condensate running down the cooled pump walls as a thin film reaches an area where the pump wall temperature is high because there is little water cooling, and thermal conductivity from the hot boiler base is predominant. (Preferably a radiation shield surrounds the lower part of the boiler wall in order to conserve heat.) Thus the more volatile constituents as well as the entrained permanent gases in the condensate tend to be released before the boiler is reached and are removed by the backing pump. The re-evaporating pump fluid in the boiler thus consists of only the less volatile components.

Many pumps have a side ejector stage (see Figure 7.3) working into the backing line; this maintains a low pressure in the region under the lower annular stage. The more volatile components returning to the boiler down the pump wall evaporate and are then pumped by the ejector stage into the backing line where they are removed by the mechanical pump. This lateral ejector nozzle also increases the critical backing pressure (see Section 7.5) and helps to prevent mechanical pump oil from reaching the diffusion pump boiler. The backing condenser arrangement prevents loss of the less volatile oil vapour constituents.

Fluid fractionation (fractionating pumps)

Most modern pumps are of the fractionating type where the more highly volatile constituents of the pump oil in the boiler are removed and the remaining components separated in proper order.

Figure 7.3 shows such a fractionating pump. The principle of operation depends on the fluid returning to the boiler being diverted inward towards the centre of the boiler via a long series of concentric channels. The jet assembly is constructed of a series of concentric tubes, having small openings, as shown, at their base. Some pumps additionally have concentric barriers in the base of the pump. As the oil flows towards the centre, the fluid in the outer portion is substantially heated, so that the high vapour pressure components are vaporized. As the fluid continues its inward journey, further heating drives off lower vapour pressure components. The various jets thus receive vapour from a specific annular region of the boiler. The jets nearest the backing region receive vapour from the outer regions of the boiler where the vapour pressure is highest, and the top-stage jet receives vapour from the

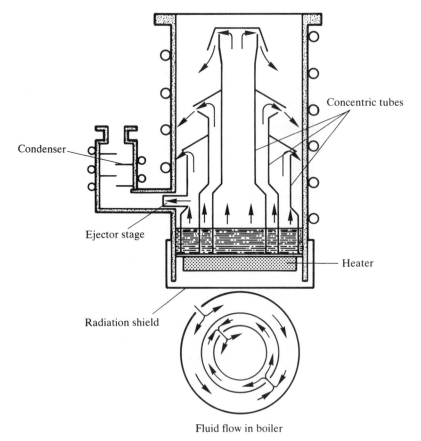

Concentric tubes

Condenser

Ejector stage

Heater

Radiation shield

Fluid flow in boiler

Figure 7.3 Fractionating diffusion pump

central region where the vapour pressure of the fluid is lowest. Since the high vacuum jet is supplied only by relatively low vapour pressure constituents, there is a reduction in the ultimate pressure obtainable compared with the same pump with a non-fractionating interior (typically one decade lower).

Figure 7.4 shows a cut-away of a diffusion pump incorporating many of the features discussed.

7.5 Critical backing pressure (CBP)

The critical backing pressure of a diffusion pump is the maximum permissible pressure allowable in the pumping line between the diffusion pump and the rotary pump. This line is known as the backing line (terminology generally used in Europe) or foreline (American terminology). If the CBP is exceeded, the normal pumping action of the jets in the diffusion pump ceases

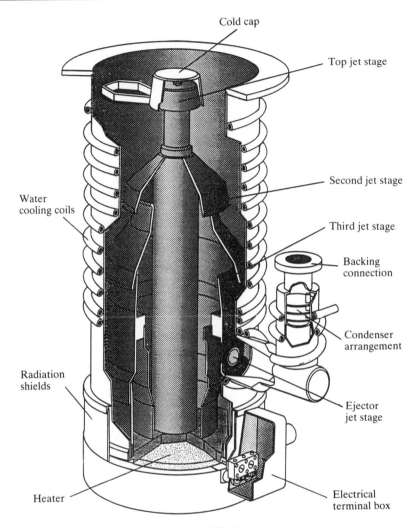

Figure 7.4 Cut-away view of a modern diffusion pump

and the pump 'stalls'. Essentially the vapour jet of the final or ejector stage of the pump is unable to sustain the pressure in the backing line when it reaches the CBP. This, in turn, causes a breakdown in sequence of the remaining jets, resulting in oil vapour being directed towards the vacuum chamber and the cessation of pumping by the diffusion pump (see Figure 7.5).

CBP is the value of backing pressure at which a slight rise in backing pressure causes a rise in the ultimate pressure on the high vacuum side.

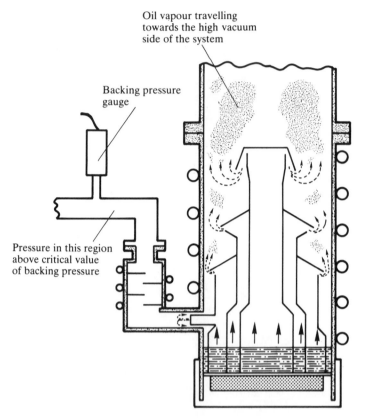

Oil vapour travelling
towards the high vacuum
side of the system

Backing pressure
gauge

Pressure in this region
above critical value
of backing pressure

Figure 7.5 A 'stalled' diffusion pump causing severe contamination of the vacuum system

The value of CBP for a particular pump is dependent on the fluid being used; for oils, this can be in the range 0.1 to 1 mbar. Values of CBP are normally quoted in the manufacturer's catalogue. Choice of rotary pump capacity is thus important since under normal working conditions it must be capable of keeping the pressure in the backing line below the CBP during conditions of maximum throughput. Ways of calculating the size of pump required will be dealt with in Chapter 13.

7.6 Diffusion pump fluids

A variety of liquids have been used as working fluids for diffusion pumps. Major factors to be considered when determining the suitability of a fluid for use in a diffusion pump are:

- Vapour pressure
- Thermal stability at boiler temperature
- Chemical inertness
- Toxicity
- Flash point
- Viscosity at room temperature
- Cost

The real test of a fluid, used in a diffusion pump, is that the pump meets the following criteria:

- Low ultimate pressure
- High pumping speed
- High critical backing pressure

The original fluid used in diffusion pumps was mercury. Unfortunately it has a number of bad features. It has a vapour pressure at room temperature of 10^{-3} mbar. The vapour pressure of the pump fluid to a great extent determines the ultimate pressure attainable by a diffusion pump. The main requirement for diffusion pump fluids therefore is that they possess a very low vapour pressure. Cold traps are necessary with mercury diffusion pumps to reach ultimate pressures lower than 10^{-3} mbar (see Section 7.7). Mercury is toxic and it amalgamates with materials such as aluminium and copper, which must therefore be excluded as pump constructional materials. The condensation of mercury on the cold walls of the pump is adversely affected by small amounts of hydrocarbon contamination within the pump; this in turn increases back-streaming and decreases the pumping speed. Despite this, mercury is still occasionally used as an operating fluid. It has advantages in systems where the use of oil might cause detrimental contamination, e.g. in the presence of hot filaments or an electrical discharge, where the oil vapour could decompose. It can be used in situations where the end product will contain mercury; thus contamination with mercury vapour from the pump would not be a problem. Note that a mercury diffusion pump will not work satisfactorily with diffusion pump oils since the interior jet gaps for oil pumps are of different sizes to those of mercury pumps.

The first low vapour pressure oils for diffusion pumps were introduced in 1929. A list of presently more popular fluids and their properties is shown in Table 7.1.

Paraffinic oils

Paraffinic oils used in diffusion pumps are either naturally occurring hydrocarbon oils which have been purified or synthetic oils of the same chemical structure. They are widely used in vacuum metallurgy and other applications in which ultimate pressures of 10^{-4} to 10^{-6} mbar range are

Table 7.1   Selected characteristics of diffusion pump oils

Chemical type	Paraffinic	Silicones	Polyphenyl ether	Perfluoropolyether*
Fluid trade names	Apiezon A, B, C Convoil Diffelen	702 . AN 120/130 704(F4) . AN140 705(F5) . AN175	Santovac 5 Convalex 10	Fomblin/Krytox
Ultimate vacuum achievable at 20 °C (typical) (mbar)	5×10^{-5} to 10^{-7}	5×10^{-6} to 10^{-9}	10^{-9}	2×10^{-8}
Oxidation resistance at boiler temperature	Poor to fair	Excellent	Very good	Excellent
Chemical resistance	Poor	Generally good	Good	Generally excellent
Cost	Low	Medium	High	High
Main applications	General purpose, mainly scientific	Industrial processing plant	Ultra-high vacuum and scientific instruments	Special-duty aggressive chemicals and oxygen pumping

* The use of Perfluoropolyether in diffusion pumps is not recommended (see text).

desired. Unfortunately these fluids have poor oxidation resistance. Exposure to air at elevated temperatures seriously degrades the oil.

Silicone oils

Silicone oils have greater resistance to oxidation when compared with other organic fluids. An example of their use is for fast cycling, valveless pumping systems used in television tube evacuation. Such applications inevitably expose the hot oil to atmospheric air. Their excellent oxidation resistance is shown in Figure 7.6. The photograph compares diffusion pump interiors, used under simulated television tube manufacturing conditions. These simulated tests repeatedly expose the hot pump oil to the atmosphere for short periods of time. The interior on the left of the photograph came from a diffusion pump using silicone oil which had undergone 1100 cycles without deterioration of the fluid. The remaining interiors using paraffinic oils had only achieved about 440 cycles on average.

Various grades of silicone are available with Silicone 705, having a vapour pressure at room temperature of about 10^{-9} mbar. One disadvantage of silicone fluids is that vapour entering a vacuum chamber (due to backstreaming) where charged particles are present, are likely to break down

Figure 7.6 Jet assemblies from different diffusion pumps that have undergone working cycles like those found in the television tube manufacturing industry

under ion bombardment (polymerization), producing insulating films. This can result in a buildup of an electrical charge on the film, which may stop the equipment functioning correctly. Silicone fluid is therefore not recommended for use in scientific instruments where energetic particles may be present.

Polyphenyl ether

The polyphenyl ethers are usually recommended to eliminate the polymerization problem in appliances such as mass spectrometers and electron microscopes. Interaction of charged particles with polyphenyl ether produces a conducting polymer film. Charges do not therefore normally build up on the film, as is the case with silicone oil. This fluid offers unusually high thermal and chemical stability with a possible ultimate pressure of 10^{-9} mbar. Originally developed as a lubricant for use in space, it was introduced as a diffusion pump fluid in the sixties.

Perfluoropolyethers (PFPE)

Like the inert PFPEs used in rotary pumps (see Section 5.4) suitable grades are available for diffusion pump use. The usefulness of these fluids lies in being inert to many of the more aggressive chemicals, among them oxygen and halogens. Applications where they have been used include ion implanters and chemical vapour deposition equipment. In general gaseous products are formed under energetic particle bombardment. If overheated (around 300 °C), decomposition forms aggressive, toxic compounds. Since such temperatures may be produced inadvertently in a diffusion pump (through misuse for example), and with the potential for toxic gas release into the atmosphere, the use of the fluid is not recommended in diffusion pumps.

Table 7.2 shows the relative cost of some of the oils discussed. An example is polyphenyl ether, which one supplier was selling in 1988 at the equivalent of about £0.8 per millilitre (when 500 ml was purchased). Care to save all usable oils is obvious during pump maintenance but is especially important when using very expensive fluids.

Table 7.2 Relative cost of some diffusion pump fluids

Diffusion pump fluid	Relative cost (1989)
Apiezon A	1 unit
Silicone 704	1.3 units
Silicone 705	2.0 units
Santovac 5	6.8 units

7.7 Baffles and traps

Back-streaming is usually quoted as the average quantity of fluid (in milligrams) reaching every square centimetre of the pump mouth area in one minute. A well designed modern diffusion pump may, for example, have a quoted value of 8×10^{-5} mg cm^{-2} min^{-1}. This represents a possible fluid loss, into the high vacuum system, for a six inch diameter pump of 0.02 cm^3 per day. Spread over a chamber surface area of, say, $2\frac{1}{2}$ m^2, this is equivalent to an oil film thickness of approximately 8 nm (80 Å) building up per day, which is an intolerably high value for many applications. However, it can be reduced by fitting a baffle between the vacuum chamber and diffusion pump inlet.

A baffle can be defined as a device consisting of a number of cooled surfaces placed near the inlet of a vapour pump to condense back-streaming vapour, condensed fluid being returned to the pump. The baffle temperature must be less than surfaces within the vacuum chamber. Within reason, the lower the baffle temperature (normally kept above the freezing point of the fluid) the more effective it becomes and the greater the reduction in chamber pressure.

A trap operating at much lower temperatures is similarly defined, except that it cannot return the condensate to the boiler. The condensing vapours collected by a trap can originate either from the diffusion pump, due to back-streaming, or from the vacuum chamber, e.g. due to outgassing. As much as 90 per cent of the outgassing load can be water vapour which is cryopumped by the trap at a high pumping speed. Figure 7.7 shows a cross-section through a baffle and trap.

The pressure in the vicinity of the trap or baffle is generally sufficiently low to produce molecular flow. Under these conditions the vapour and gas molecules travel in straight lines between collisions with surfaces or with each other. The trap or baffle is designed with a minimum requirement that no straight-travelling molecule can pass through the trap or baffle without undergoing at least one collision with a cooled surface.

For many applications of oil diffusion pumps the water-cooled baffle system is sufficient to maintain the base pressure and surface cleanliness required. This will depend on the fluid used. Pressure readings of 10^{-6} mbar are typically maintained in systems using Silicone 702 and in the low 10^{-9} mbar range in systems employing Santovac 5. Refrigerating the baffle (but not below the freezing point) will result in a lowering of the vapour pressure of the oil so that base pressures about one decade lower are possible.

To realize base pressures significantly less than this requires the use of traps at much lower temperatures, e.g. liquid nitrogen temperatures ($-196\,^{\circ}$C), since some of the light fractions from decomposed pump oil cannot be condensed at refrigerated baffle temperatures.

Design requirements for such low-temperature types of trap are a low refrigerant consumption rate and a sufficiently large reservoir to accommo-

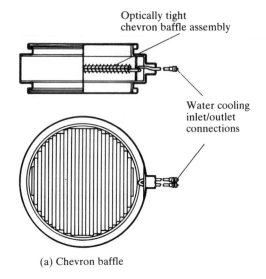

(a) Chevron baffle

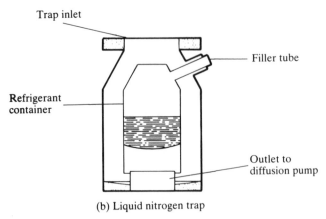

(b) Liquid nitrogen trap

Figure 7.7   Cross-section through a baffle and trap

date an overnight charge; otherwise the provision of an automatic topping-up device is required.

When a liquid nitrogen trap is used directly above a diffusion pump, the pump fluid will condense and freeze on the trap. In the extreme, under continuous operating conditions over long periods of time, the pump may lose most of its fluid, to be frozen on the trap. The solution of this problem is to interpose a cooled baffle between the pump and the liquid nitrogen trap; this suppresses the majority of the back-streaming fluid and allows it to return to the boiler.

Combinations of traps and baffles have been quoted as achieving reductions

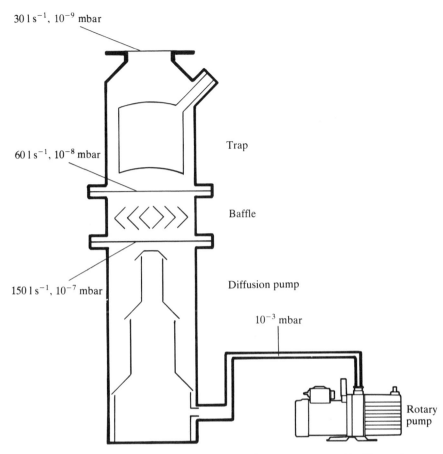

$30 \, 1 \, s^{-1}, 10^{-9}$ mbar

$60 \, 1 \, s^{-1}, 10^{-8}$ mbar

Trap

Baffle

$150 \, 1 \, s^{-1}, 10^{-7}$ mbar

Diffusion pump

10^{-3} mbar

Rotary pump

Figure 7.8   Typical effect on pump speed and ultimate pressure of baffles and traps

in back-streaming approaching 800 times, giving a value of approximately 10^{-7} mg cm^{-2} min^{-1}. Such extremely low rates are negligible for most applications. The main contributor of contamination is now the system itself.

Figure 7.8 shows how combinations of baffles and traps affect pumping speed and ultimate pressure. Note that the effective pumping speed above the liquid nitrogen trap is about one-fifth of that of the diffusion pump. To obtain pressures below 10^{-7} mbar all elastomer seals must be replaced with metal gaskets because of the increasing influence of outgassing of materials at low pressures. The contribution due to outgassing of the walls and materials within the system itself can be reduced by careful selection of system materials, by chamber design to minimize surface areas and by system baking to accelerate desorption. A more detailed discussion of material selection and baking will be dealt with in Chapter 13.

7.8 Diffusion pump system operation

Let us consider how we would operate the valveless diffusion pump system shown in Figure 7.9a.

1. *Initial condition of system.* The rotary pump is switched off. The diffusion pump heater is off. The air admittance valve is closed. All gauges are off and the entire system is at atmospheric pressure.

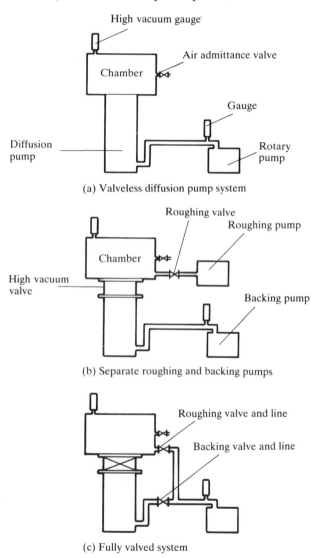

(a) Valveless diffusion pump system

(b) Separate roughing and backing pumps

(c) Fully valved system

Figure 7.9 Different diffusion pumped system layouts. Note that the backing pump is sometimes referred to as the forepump (American terminology).

2. *Pump-down of system.* Switch the rotary pump on and check the pressure in the backing line; this must be less than 0.1 mbar. Turn on the water and diffusion pump heater; when the diffusion pump is operating (after about 20 minutes) switch on the high vacuum gauge. The pressure in the chamber should be in the high vacuum region.

The real disadvantage of the system comes if we wish to let the chamber up to atmospheric pressure. The only option would be to switch off the high vacuum gauge, turn off the diffusion pump heater and wait for the oil in the diffusion pump to cool down to a temperature where air admission will not cause its decomposition. Note that during the cool-down period (which may be at least 20 minutes), the rotary pump would be left running (backing). On cyclic operating systems where a process demands frequent operation between atmospheric pressure and high vacuum this waiting period between oil warm-up and cool-down cannot be tolerated. Additionally, back-streaming from the diffusion pump can be significantly high during these transient periods.

A better solution is shown in Figure 7.9b. Here a high vacuum isolation valve (HVV) has been added above the diffusion pump and another rotary pump (roughing pump) and roughing valve attached to the chamber. New operation would be as follows:

1. All valves closed. Backing pump on, check pressure in backing line, which must be less than 0.1 mbar, water on, heater on, wait for oil to boil. During this waiting period the chamber can be 'roughed out' by switching on the roughing pump and opening the roughing valve. Pump the chamber down to less than 0.1 mbar. (Do not leave the roughing pump on for too long at or near the pump's ultimate pressure, since contamination of the chamber may occur through back-migration of rotary pump oil vapour.)
2. Once the chamber has been roughed down to less than 0.1 mbar it is best to close the roughing valve to isolate it from the roughing pump.
3. When the diffusion pump is operating and the roughing valve is closed, open the HVV; the pressure in the chamber should drop to high vacuum. Now if we wish to 'vent' the chamber to atmospheric pressure the diffusion pump can be left running but isolated from the chamber by closing the HVV and opening the air admittance valve. Pumping out the chamber now would be a matter of closing the air admittance valve, 'roughing out' the chamber again with the roughing pump, isolating the roughing pump and opening the HVV. A much faster cycle time is thus achieved. The main disadvantage of this system is the additional cost of the roughing pump.

A cheaper solution, shown in Figure 7.9c, uses a single rotary pump for both backing and roughing duties (but only performing one function at a time).

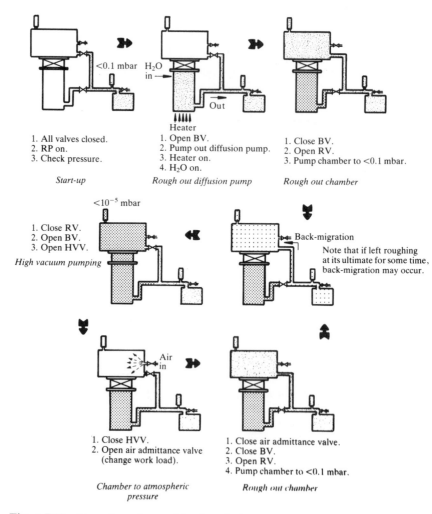

1. All valves closed.
2. RP on.
3. Check pressure.

Start-up

<0.1 mbar

1. Open BV.
2. Pump out diffusion pump.
3. Heater on.
4. H_2O on.

Rough out diffusion pump

Heater
H_2O in →
Out

1. Close BV.
2. Open RV.
3. Pump chamber to <0.1 mbar.

Rough out chamber

$<10^{-5}$ mbar

1. Close RV.
2. Open BV.
3. Open HVV.

High vacuum pumping

Back-migration

Note that if left roughing
at its ultimate for some time,
back-migration may occur.

1. Close HVV.
2. Open air admittance valve
(change work load).

Air in

*Chamber to atmospheric
pressure*

1. Close air admittance valve.
2. Close BV.
3. Open RV.
4. Pump chamber to <0.1 mbar.

Rough out chamber

Figure 7.10 Operation of a combined outfit (start at top left-hand corner)

A comprehensive description of the start-up, cyclic operation and shut-down procedure for a fully valved system (combined outfit) is given below. Figure 7.10 shows the various steps in the operation.

Start-up

Switch on the mechanical pump and with all the valves closed the performance of the mechanical pump is checked. It should produce less than 0.1 mbar if in good condition.

Rough out diffusion pump

Open backing valve. Pump-down the diffusion pump via the backing line to less than 0.1 mbar.

It may be advisable at this stage to check the leak-tightness of the vacuum chamber and roughing line before switching on the diffusion pump. Obviously this is impracticable if the chamber is large and long roughing times are expected. If this is the case then the diffusion pump heater and water supply can at this stage be turned on.

Rough out chamber

Close backing valve. Open roughing valve and rough out the system to 0.1 mbar or below. Remember it is inadvisable to leave the system roughing at low pressures for long periods of time because of the possibility of oil vapour back-migration from the rotary pump into the chamber.

Isolate the chamber under vacuum and return to backing the diffusion pump by closing the roughing valve and opening the backing valve.

High vacuum pumping

When the diffusion pump is operational shut the backing valve and open the roughing valve and check that the chamber pressure is still below 0.1 mbar. Then close the roughing valve, open the backing valve and slowly open the high vacuum valve. The pressure in the chamber should now drop rapidly, as indicated by the high vacuum gauge. Valving-in at pressures higher than 0.1 mbar in the chamber may possibly cause the pump to momentarily 'stall'. Additionally, for similar reasons, under high vacuum pumping it is imperative that the air admittance valve should not be opened so that a rise in pressure occurs sufficient to 'stall' the diffusion pump; otherwise diffusion pump vapour is likely to move into the chamber and mechanical pump. The onset of 'stalling' usually occurs at the 'critical backing pressure' (as discussed in Section 7.5).

> The most important rule of diffusion pump operation is that the critical backing pressure should *not* be exceeded under any circumstances.

Cycling the chamber between atmospheric pressure and high vacuum

1. *Chamber to atmospheric pressure.* With the chamber at high vacuum close the high vacuum valve. Switch off the high vacuum gauge and open the air admittance valve. In an industrial process, the chamber may have a door;

this could now be opened and the work load inside removed and replaced. Shut the chamber door and close the air admittance valve.

2. *Rough out chamber.* Close the backing valve, open the roughing valve and rough out the chamber.

3. *High vacuum pumping.* With the chamber pressure below 0.1 mbar, the roughing valve is closed and the backing valve opened. The high vacuum valve is now opened and the chamber pressure should quickly return to high vacuum.

One cycle has now been completed.

Shut-down

1. Switch off all electrical gauges.
2. Close the high vacuum valve, thus isolating the chamber under vacuum.
3. Switch off the diffusion pump and allow it to cool for at least 20 minutes. Do not leave in this condition for too long as rotary pump oil vapour may back-migrate into the cold diffusion pump.
4. Close the backing valve, isolating the diffusion pump under vacuum.
5. Switch off the mechanical pump and admit air to it. Turn off the water.

Notes

1. As an alternative it is permissible (providing the valves are leak-tight), at step 3 above, to switch off the diffusion pump and to isolate immediately the diffusion pump by closing the backing valve and letting it cool down under vacuum.

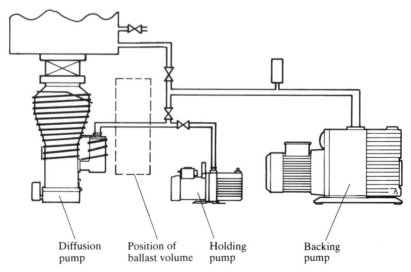

| Diffusion pump | Position of ballast volume | Holding pump | Backing pump |

Figure 7.11 Use of a holding pump (or alternative ballast volume) where roughing times are long

2. Backing and roughing valves are not normally operated so that they are both open at the same time, since this effectively 'short circuits' the diffusion pump. Usually one is closed before the other is open.
3. If valves are of the diaphragm type, do not close excessively tight or the diaphragm may become damaged and cause leakage across the valve seat.
4. On some systems roughing times may be long, and there is a danger of the backing pressure on the isolated diffusion pump side of the closed-off backing valve exceeding the critical backing pressure. A solution in such cases is to increase the volume of the backing line, i.e. by adding a tank or 'ballast volume' as it is known. This will allow the pump to be isolated for longer periods before the critical pressure is reached. Another alternative is to add an additional small rotary pump (a 'holding pump') and valve, built into the system between the diffusion pump and the backing valve (see Figure 7.11). The size of such holding pumps is about 10 per cent of the normal backing pump displacement.

7.9 Integrated vapour pumping groups

A vapour pumping group arrangement of pump/baffle/trap components would normally be designed as separate entities and would be bolted together to provide the required grouping. We have seen that the effective pumping speed at the inlet of such a group is much lower than that of the pump itself, due to the poor conductance of the assembly. At least three sealing gaskets, together with a number of associated flanges, would be required. Such a combination is heavy, bulky and can be complicated to install. Additionally, each elastomer seal contributes to the outgassing load, which imposes a limit to the possible ultimate pressure attainable. There may also be trapped volumes of gas involved which can lead to gas pressure bursts.

There are thus important benefits to be gained by integrating the pump and accessories within a single housing. Examples of such units are those marketed by Edwards High Vacuum International and known as 'Diffstak'; 'Crystal' by Alcatel; or the 'Diffcompac' range by Leybold. In its simplest form it combines a three-stage diffusion pump and an optically opaque water-cooled baffle all integrated in a single casing. Other versions additionally combine a quarter-swing valve and roughing line connection. Such single-structure pumping groups are faster, cleaner and smaller than an equivalent modular system. Figure 7.12 shows an integrated design contrasted with a conventional pumping group of similar performance, but assembled from separate components. The saving in space occupied is obvious. The single-structure group simply needs to be bolted to the vacuum system and to be provided with a line to the backing pump; the vacuum circuit is then complete.

The integrated design has meant a significant reduction in the number of

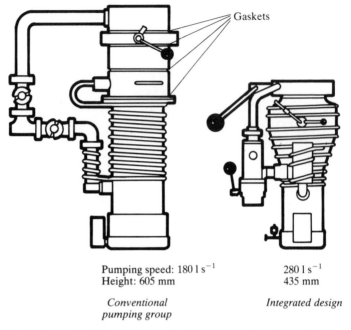

Pumping speed: 180 l s^{-1} 280 l s^{-1}
Height: 605 mm   435 mm

Conventional   *Integrated design*
pumping group

Figure 7.12 Integrated design contrasted with a conventional pumping group

seals; this reduction in the use of large 'O' rings and the associated crevices has proved to be very important in achieving an ultra-clean vacuum.

In certain applications, it is useful to incorporate a liquid nitrogen cooled surface in addition to the water-cooled baffle. These pumps have an extra space between the water-cooled baffle and the high vacuum valve into which is fitted a baffle structure made of copper. This is cooled by thermal conduction from an external liquid nitrogen reservoir. Figure 7.13 shows a cross-section through a cryo-cooled version of the integrated design.

7.10 Vapour booster pumps

Vapour booster pumps have evolved from conventional diffusion pumps and are designed to have high pumping speed and high gas throughput in the range 0.5 to 10^{-4} mbar. Additionally, they are capable of delivering gas against a backing pressure of a few millibars. They need to be operated in series with a mechanical pumping arrangement capable of delivering to atmosphere. Typically this could be either a single- or a two-stage rotary pump, or a mechanical booster pump/rotary pump combination.

Figure 7.14 shows a schematic diagram of a vapour booster. The essential parts, like the diffusion pump, are a water-cooled body, an interior assembly containing the jets and a boiler for evaporating the working fluid.

Conduction-cooled
copper rod

Liquid nitrogen
reservoir

Liquid nitrogen
cooled disc

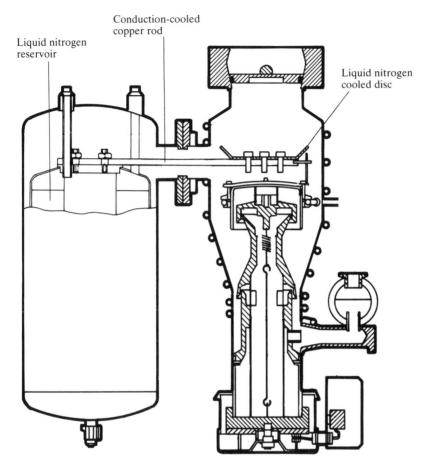

Figure 7.13   Cryo-cooled version of the integrated design

Pump fluid is heated in the boiler by the heaters. The vapour generated passes through the vapour tubes and out of the annular jets and ejector nozzle as a high-velocity jet of vapour. The vapour jet drives the gas through the pump by entrainment. The vapour condenses on the water-cooled cone-shaped casing walls and returns to the boiler via fluid return tubes. Vapour that fails to condense on the casing walls passes into the backing condenser where it has a further opportunity to condense. A cold cap is fitted as standard to minimize back-streaming. A dipstick is provided to check the fluid level, since fluid loss is generally greater for vapour boosters than for diffusion pumps.

Because boosters are required to cope with high gas throughput in the medium vacuum range, heater powers are greater than corresponding sizes of diffusion pump. Fluid capacities are also larger, to allow for the significant

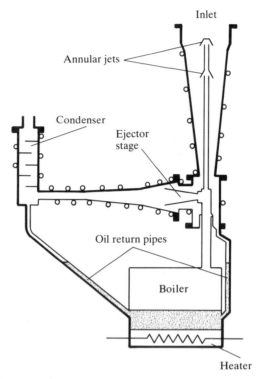

Figure 7.14   Oil vapour booster pump

losses of fluid into the backing line which can occur during operation. Boilers thus tend to be comparatively large. Efforts are made to minimize heat loss from the boiler (e.g. using lagging and/or radiation shields).

Fluids used in vapour boosters need to be considerably more volatile than those used in diffusion pumps, in order to generate the very powerful vapour jets required without running boilers at temperatures likely to cause fluid decomposition. Typically, booster boiler pressures are 25 to 50 mbar compared with 2 to 5 mbar for diffusion pumps. Fluids must also have a vapour pressure at 20 °C which is sufficiently low to give an ultimate pressure of less than 10^{-4} mbar.

Vapour boosters are employed in plant used for the following processes:

1. Vacuum metallurgy—refining, heat treatment, alloying, degassing and brazing of metals
2. Vacuum drying, impregnation, degassing and distillation
3. Vacuum metallizing, particularly when plastics or paper are coated
4. Low-density wind tunnels

In furnace applications, dust and grit may enter the pumps and find its way into the boiler return line and the boiler itself. In time this may lead to excessive fluid loss from the pump, causing the boiler to become red hot and in the extreme to collapse inwards under the action of atmospheric pressure. Accessories such as inlet dust filters and thermal switches are useful in such applications. Some boosters have boiler access plates to allow removal of 'sludge' from the boiler or fluid return lines during maintenance.

Also during maintenance nozzle throats should be checked for blockage and vapour chimneys and nozzles should be clean inside and out.

7.11 Looking after diffusion pumps

Changing fluids

Most diffusion pump fluids (except perfluoropolyether) are compatible and could be mixed with each other. In general, however, mixing of fluids is not recommended since, for example, if various grades of silicone fluid were combined the ultimate pressure obtained would be determined by the higher vapour pressure constituent of the mixture. Additionally, for example when replacing a silicone fluid (DC 704) with polyphenyl ether, which has a high boiling point, silicone fluid left in the pump can break down at the high temperatures to give tar and hydrogen. When changing from a high boiling point fluid to a low boiling point fluid, a small quantity left in the pump will probably have no effect.

Some users who replace fluids on a regular scheduled maintenance will consider reclaiming contaminated or thermally degraded oils. Obviously the economics of reprocessing the used fluids as against replacing them must be taken into account.

Diffusion pump maintenance

Diffusion pumps on systems that are infrequently cycled between vacuum and atmospheric pressure generally require little maintenance. Usual problems, apart from the case of misuse, are failure of the heaters or interruption of the water or air cooling.

In a pump operated deliberately or inadvertently at or near atmospheric pressure for some time, the temperature of the fluid rises and decomposition takes place. In such cases eventual disassembly of the pump will be necessary, since all diffusion pump fluids, except perfluoropolyether, reduce to a carbon-like deposit.

Diffusion pumps are usually designed to be readily dismantled for cleaning, in accordance with the manufacturer's instructions. There is a temptation to remove and work on a pump while it is still warm. For health reasons care must be taken never to inhale vapours or mists emitted from a pump. Allow it to cool thoroughly. The only cleaning procedure for decomposed fluid deposits is mechanical abrasion, done by hand on open portions of the pump and by a shot blaster in inaccessible parts. All surfaces of the pump which show tar-like deposits may be cleaned with acetone or trichloroethane and a stiff brush or a scouring pad which is not impregnated with cleaning agents. Occasional difficult spots may be rubbed clean with a fine grade of emery cloth. Care must be obviously taken with plated items. All traces of emery must be flushed away with solvent and component parts fully dried. If the jet assembly has been dismantled, care should be taken to ensure that all parts fit together correctly.

With regard to shot-blasting, for steel jet assemblies a nickel shot is ideal; for aluminium, glass beads are recommended. Although both dry and wet beads may be used, dry beads are preferred, mainly because additives which may be used in wet bead blasting could lead to contamination.

Table 7.3 gives a check-list of points to look for when dismantling/inspecting and reassembling diffusion pumps. Points to note are that interior parts do not have to be highly polished. Jet gap widths must be the same all around a particular stage; closure due to damage of a jet cone must be rectified. 'O' rings used to seal the dipstick or drain, etc., often become hard, brittle or have a permanent 'set' due to the high-temperature conditions; these must be replaced. The direction of the cooling water flow should be down the pump from the high-vacuum inlet towards the backing connection. The pump performance is particularly sensitive to the cooled wall temperature around the top pumping stage.

Table 7.3 Dismantling/inspecting/reassembling diffusion pumps—a check-list

1. Inspect interior parts for cleanliness.
2. Jet gaps—check concentricity.
3. Check elastomer seals for permanent set/hardness, etc.
4. Is cooling water flow sufficient or is it restricted by deposits of calcium salts?
5. Check fluid condition and quantity.
6. Are heaters in good thermal contact? Check flatness and tightness.
7. Inspect jet assembly orientation—ejector stage must point towards the backing connection.
8. Inspect setting and condition of cold cap.

Figure 7.15 illustrates how the ultimate pressure obtained, using some of the pump oils, varied with water temperature for one particular pump and water-cooled baffle combination. Pressures were measured above the baffle with an ionization gauge; metal seals were used. For some of these fluids it can be seen that a 15 °C drop in incoming water temperature lowers the ultimate pressure achieved by about a decade.

Cooling coils for large pumps may be split into two or three sections. Avoid starting the pump without proper cooling. On some pumps that have a soft solder fillet between the cooling coil and body, insufficient cooling can cause

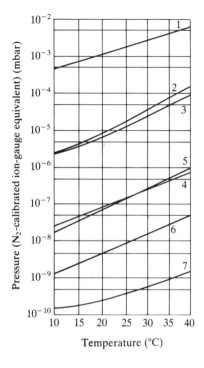

KEY

1 Mercury
2 Silicone 702
3 Apiezon B
4 Apiezon C
5 Silicone 704
6 Silicone 705
7 Santovac 5

Figure 7.15   Variation of ultimate pressure with pump and baffle cooling-water inlet temperature for various fluids used in a diffusion pump

the soft solder to melt, run down the pump and most likely collect on the floor! This will upset the cooling efficiency. Without adequate cooling, the pump fluid vapours will not condense and may migrate throughout the system, causing contamination and fluid loss. With correct water flow rates, the water temperature rises in most cases by about 10 to 15 °C.

Depending on the water supply, 'furring' of the pipes due to calcium salts may occur over a long period of time. These can be removed with an industrial descaler. Care must be taken to ensure that such descalers are compatible with the pipe material. Additionally, these materials are often strongly acidic and contact with skin, eyes or clothing must be avoided. Multiple heaters are often fitted to large pumps and it is possible for one of these to fail and result in reduced heating. If possible check with a wattmeter.

Where heaters are secured to a pump base by retaining studs it is useful to lightly coat the threads of the stud with an antiseize compound before fitting the securing nut. This will ensure easy subsequent removal. Poor clamping with inadequate thermal contact may result in reduced heater life. Clamping bolts should be tight enough to avoid air gaps between heaters and the boiler plate.

New diffusion pump oil usually gives off large quantities of dissolved air when it is first heated. This gas evolution may lead to a noticeable high backing pressure or give pressure bursts, and should not be mistaken for a leak.

7.12 Troubleshooting a diffusion pumped system

When a problem occurs in a vacuum system, it is best to approach the situation in a systematic manner, rather than to 'jump in' and try to resolve the problem by any means that may occur to us at random.

To assess the degree of system malfunction, we must first know what may normally be expected of it. System pressure and time to reach vacuum levels could be recorded. Referring to Figure 7.16, which represents a typical high-vacuum system, pressure readings should be available at the chamber, in the backing line and in this case in the holding pump line. If a record of pressure readings has been kept we can compare its performance with the recorded information.

Readings should be taken at comparable times during a typical routine pumping cycle, and not for a system that has just been started up after routine shut-down on recent cleaning. Under such conditions, it may take 24 to 48 hours before routine operational conditions are again obtained.

Checking the rotary pump and pumping lines

Let us assume that the initial procedure for the operation of this particular system would be to:

1. Use the main rotary pump to evacuate the chamber to below 0.1 mbar; i.e. backing valve closed, roughing valve open, high vacuum valve closed.
2. At the same time the holding pump is used to evacuate the diffusion pump and get the diffusion pump heated up; i.e. holding valve open.

In the first instance, let us suppose that it is not possible to pump the chamber via the roughing line to below 0.5 mbar (as measured by the chamber gauge). First we must recognize that the gauges themselves can malfunction, either failing completely or giving false information as to the system's condition. Failure to recognize this fact can lead to a lot of wasted effort. A quick check is available by comparing the chamber pressure reading with the backing line pressure. If they both read approximately the same it

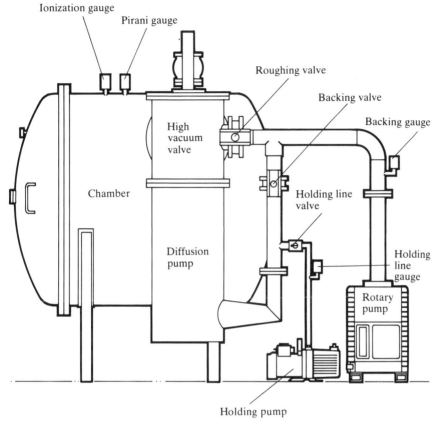

Figure 7.16 Layout of the high vacuum system under investigation in Section 7.12

can be concluded that the chamber gauge is functioning properly. If the backing line pressure is substantially different, the chamber gauge may be at fault. Try replacing it or swapping it for the backing gauge.

If the indicated chamber pressure appears to be correct try isolating the rotary pump from the system by closing the roughing valve and check its performance. If the pressure remains high compared with normal, there are two possibilities: either the rotary pump is not operating properly or there is a relatively large leak in the pumping lines up to the backing and roughing valves.

An open gas ballast valve makes it impossible to achieve normal non-ballasted ultimate pressure. The gas ballast valve may have been opened to prevent water vapour condensation or to clear the pump of water contamination. It may then have been left open unintentionally. This can often happen in a multiple-shift operation: someone on one shift may open the gas ballast valve and the next shift may be unaware that it was left open.

In the case of belt drive pumps, poor ultimate pressure can result from a loose drive belt. Belt tension can be adjusted. Frayed belts should be replaced. With pumps with multiple belts, all belts should be replaced together.

The most common cause of poor vacuum associated with the rotary pump is contaminated oil. If there is any indication of discoloration, low operating level or thinning out of the oil, drain and flush (with fresh oil) and refill. Switch on the pump and allow to run for some time, to allow the fresh oil to degas and the pump to reach operating temperature. If this fails to improve the ultimate pressure reached, it would indicate possibly a more serious problem with the pump or a leak.

The next step would be to disconnect the pump from the backing line and to connect a gauge directly to the pump inlet. If the pump pressure indicated is reduced to a lower level, it can be assumed that the problem is a leak in the pumping lines. Inspect joints to see if they have worked loose through vibration. In some systems, vacuum hoses are used in these lines. These may deteriorate with time. If a leak is not obvious, leak detection must be carried out (see Chapter 14). Leaks certainly are a major reason for system malfunctions. Air leaks are not uncommon. It is worth determining if any maintenance work has been done on the system, since a seal that has been broken and remade can inadvertently become a source of leakage, if not carried out satisfactorily. Assuming a leak is found and repaired, normal system performance should be obtained.

It should be borne in mind that valves can malfunction and although we are leak-testing the pumping lines up to the valves it is possible that in this case perhaps the roughing valve or backing line valve is leaking. Wear, caused by high friction forces or debris, can prevent proper sealing. The seals, bellows, etc., can wear in time. The actuating air pressure on pneumatic valves may be too low to provide proper sealing. Solenoids may be faulty, etc.

Checking the diffusion pump

With the holding valve closed, check the holding line pressure (assuming this gauge is functioning satisfactorily); if the pressure is substantially above 0.05 mbar it may well be worth checking the pump and checking for leaks as above. Let us assume that the pressure is below 0.03 mbar, which is acceptable. Open the holding valve and pump out the diffusion pump cold (with backing valve closed). Check the pressure again; if the diffusion pump was left under vacuum the pressure should fall rapidly to, say, 0.05 mbar.

Note that when evacuating the diffusion pump from atmospheric pressure it will be quicker to use the main rotary pump via the backing valve. Pressure bursts may be due to outgassing of dissolved air in the oil; also in some situations it has been known for water vapour from the air in the diffusion pump to have condensed on the cold walls of the pump while it has been standing at atmospheric pressure. Water-cooled baffles and cold caps have been known to leak water into the diffusion pump. Water leaked into a vacuum system, even on a tiny scale, is a serious problem and one often hard to resolve. It is well to remember that a few drops of liquid represent a substantial amount of vapour when evaporated. Another problem encountered has been leaks through diffusion pump drain seals, sight glass seals, dipstick seals, etc.

Assuming that the pressure in the diffusion pump is below or around 0.05 mbar, the heaters can be switched on (check that water is flowing) and the diffusion pump allowed to heat up, backing preferably with the main rotary pump. Again pressure bursts may be observed as the oil is heated.

With the diffusion pump working normally and the high vacuum valve open, a slow pump-down is normally associated with a high gas load. This is evidenced by high backing line pressures. High gas load can come from two sources, leakage and outgassing. Records of 'normal' operation will give a clue to whether this is acceptable.

When a problem is definitely indicated, try pumping-down twice. Plot the pressure versus time curve and note the time (t_1) between two arbitrarily selected valves of pressure, say P_1 and P_2 (see Figure 7.17). Close the high vacuum valve and allow the pressure to rise naturally to the pressure P_1. Open the high vacuum valve and record the fall in pressure again.

How long did the second pump-down take? If the time interval t_2 is approximately equal to time t_1, this would indicate that the problem might be due to an air leak. This is because the source of gas is constant, indicating an unvarying physical leak.

If the test resulted in time t_2 being about $\frac{1}{3}$ to $\frac{1}{5}$ the pump-down period of time t_1, then this probably indicates a virtual leak or an outgassing problem. The reason for this is that during the first pump-down period, much of the gas load, including the virtual leak and outgassing, was removed. Then, during

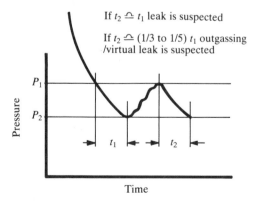

Figure 7.17   Pump-down test to determine the difference between a leak and outgassing

the second pump-down, the pump did not have as large a load to handle and therefore was able to pump-down considerably faster.

In some coating systems a gradual increase in the outgassing rate is expected as coating accumulates in the chamber. At some point this gas load becomes intolerable and a general clean-up is required. Severe outgassing may be caused by the processing of unusually 'dirty' work.

If the unit is shut down for cleaning, it may be advisable to also clean the diffusion pump and change the oil. However, one might consider the length of time since the fluid was last replaced and whether the vacuum system has been inadvertently exposed to atmosphere while hot in the intervening period. The system will stand a great deal of abuse in this regard. Units have been known to be 'dumped to air' many times before their pumping capabilities are impaired. However, if the unit has been in operation for six months or a year and situations have been moderately adverse, it is considered good practice to change the diffusion pump oil.

After the above procedures have been performed and the unit reassembled, the unit should be allowed to run for at least 24 hours. If the diffusion pump fluid has been changed, several days of operation may be necessary for the fluid to become conditioned and performance to reach its best. This is common to all diffusion pump vacuum systems.

Table 7.4 itemizes a number of possible fault situations together with probable causes and correction procedures.

Table 7.4   Fault diagnosis and correction

Fault	Probable cause	Correction
Poor ultimate pressure	1. Faulty gauge or gauge head	Check with replacement
	2. Leak in system, virtual or real	Locate and rectify
	3. System dirty	Clean system
	4. Contaminated pump fluid	Examine and renew if necessary (does it look discoloured?)
	5. Low heat input	Check heater voltage, check for continuity, burned out element, poor thermal contact
	6. Insufficient cooling water	Check water pressure, check tubing for obstructions and back-pressure
	7. High backing pressure	Check for leak in backing line, poor mechancial pump performance, breakdown of pump fluid
	8. High vacuum valve not fully open	Open fully

Table 7.4 (*continued*)

Fault	Probable cause	Correction
Low speed (prolonged cycle)	1. Low, or in the extreme, no heat input	Check power, heater, circuit breaker, thermal snap switch
	2. Low fluid level	Check and top up fluid, if necessary; do not check while pump is hot
	3. Malfunctioning pump assembly, improperly located jets, damaged jet system	Check and rectify
	4. High gas loads	If confirmed by high backing pressure, locate source; is it leakage or outgassing?
Inlet pressure surges	1. Incorrect heater input	Check and correct
	2. Fluid outgassing	Allow fluid time to degas
	3. Leak in system ahead of pump inlet	Check and correct
	4. Trapped volume in system	Check and correct
	5. Excessively lubricated 'O' ring seals	Check and correct
	6. Low fluid level	Check and top up if necessary
High chamber contamination	1. High backing pressure	Check for leak in backing line; poor mechanical pump performance; breakdown of pump fluid
	2. Prolonged operation at high throughput at pressures above 10^{-3} mbar	Review procedures
	3. Incorrect system operation and air release procedures	Review procedures

8

Turbomolecular pumps

8.1 How turbomolecular pumps work

A turbomolecular pump is a gas transfer pump which operates like an axial flow compressor used on a jet engine. Sets of moving blades, separated by stationary blades, rotate at high speed (up to $60\,000$ rev min^{-1}), receiving and compressing gas from a high vacuum chamber and delivering it to a rotary pump on the outlet side. Turbomolecular pumps are designed to operate under molecular flow conditions.

The physical basis for the pumping action is the interaction effect between a molecule and a moving surface. The short but finite residence time occurring when a molecule strikes a surface results in the molecule acquiring an additional velocity component in the direction of the moving surface (see Figure 8.1).

The orientation of the moving (rotor) blades and stationary (stator) blades to the axial direction in a turbomolecular pump is as shown in Figure 8.2. The molecule shown incident on a rotor blade will reside on the surface for a short time and then will probably leave in the direction shown due to the lateral movement of the blade and its pitch. The stator blades are also pitched. Their pitch direction is such that they preferentially transmit molecules that have left the rotors and are moving axially down the pump. Molecules moving in the reverse direction (back-diffusion) are likely to be reflected back (as shown). The other major effect of the stators is to stop the sidewards movement of molecules that have been struck by the rotors, directing the molecule velocities further into the axial direction down the pump. It is rather difficult to explain the pumping mechanism of a turbomolecular pump diagrammatically because of the high rotational speed involved. However, as

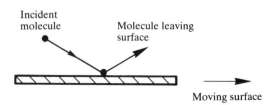

Figure 8.1   Interaction between a moving surface and an incident molecule

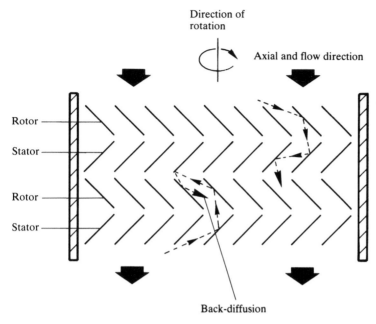

Figure 8.2 Orientation of rotors and stators in a turbomolecular pump

a result it is more probable that the molecules will be driven towards the direction of the exhaust of the pump rather than towards the inlet.

For the blades to be effective with rapidly moving molecules, the blade speeds must approach the molecule speeds; otherwise the molecules will pass through the rotor regions without being struck.

For a rotor travelling at 60 000 rev min^{-1} and with a mean blade diameter of 7.5 cm, its blade tip speed is around 236 m s^{-1} (527 miles h^{-1}). By comparison, the average speed for nitrogen molecules at 20 °C is 470 m s^{-1}. Heavier molecules will be slower, lighter molecules faster, such as hydrogen at 1900 m s^{-1}. Thus lighter molecules are more difficult for turbomolecular pumps to pump, because they are more likely to pass through the rotors without being hit by a blade, increasing the likelihood of back-diffusion of the gas.

Pump performance depends on blade design as well as rotor speed. Variables include the pitch angle, blade width and the distance between blades. Overall performance is optimized by varying the blade geometry through the pump. Most modern turbomolecular pump designs use a low compression, high pumping speed, open blade structure at the intake end of the pump and a high compression, low pumping speed, closed or overlapping blade configuration at the outlet end. This combination provides a good pumping speed and overall compression ratio.

Compression ratio

The compression ratio for a turbomolecular pump is the ratio of outlet to inlet pressures for a particular gas species when the pump is operating. Compression ratio depends on the molecular weight of the gas being pumped, typical values being 1000 for hydrogen and 10^9 for nitrogen.

Each rotor/stator combination is a compressor stage, and the total compression ratio for the pump is the product of the compressions achieved across each stage.

The compression ratios achievable by a pump for different gases are independent of pressure in the molecular flow regime. For most gases, the compression ratio begins to fall off above 10^{-2} mbar and for all gases drops to almost zero under viscous flow conditions (see Figure 8.3).

Use of a two-stage rotary backing pump to give backing pressures of better than 10^{-2} mbar is essential if low molecular weight gases are to be pumped.

8.2 Constructional and mechanical aspects

Two configurations of turbomolecular pump are produced: dual flow, where the axial flow direction is horizontal, gas enters at the mid point of the axis and flows outward towards the ends (see Figure 8.4), and single flow, where gas enters at one end via a wide flange connection and exits at the other end to the rotary pump (see Figure 8.5).

The disadvantage of the dual-flow design is the reduction of gas access to the rotors caused by the type of entrance, a right-angled entrance partially blocked by the rotor shaft. This problem has been partially offset by reduction of the shaft diameter and widening of the entrance region. One advantage of the design is that the centre section of the body including the inlet is rotatable through 360°.

Oil-lubricated ball-bearings have been used continuously since early pump designs were introduced. Entry of vapour from the lubricating oil to the vacuum during operation is not a problem because the high molecular weight of oil molecules means that the pump will exclude them most effectively. A significant amount of heat is generated and some form of cooling is required, both water and air cooling systems being available. Air-cooled pumps have the advantage of portability. However, the advisability of using air cooling depends upon the application. Air cooling will always be less efficient than water cooling. Therefore it is not desirable to use air cooling in an application with continuous high gas throughput since this generates additional heat in the pump. Applications involving frequent cycling of the pump between atmosphere and vacuum are also less suitable for air cooling since more heat is generated in the acceleration phase than when the pump is running at constant speed. In areas of high ambient temperature water cooling is preferable to air cooling.

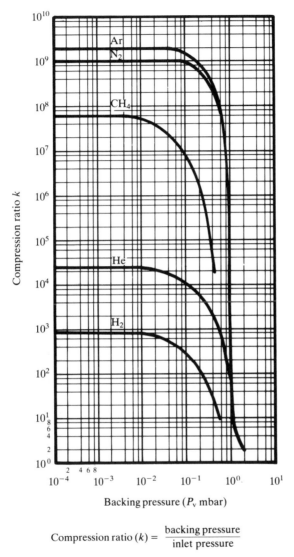

$$\text{Compression ratio } (k) = \frac{\text{backing pressure}}{\text{inlet pressure}}$$

Figure 8.3   Compression ratio curves for different gases

Grease-lubricated bearings have been available for some years. Grease-lubricated bearings can normally allow pumps to be operated in any orientation. Oil-lubricated bearings of the vertical type are limited in this respect.

Rotors are driven by a medium-frequency three-phase squirrel-cage motor directly coupled to the rotor shaft. The motor is powered from a variable-frequency controller which controls the motor speed during start-up. A

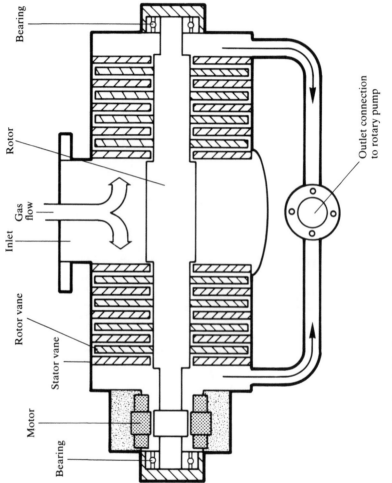

Figure 8.4 Horizontal turbomolecular pump

Bearing

Rotor

Inlet

Gas flow

Rotor vane

Stator vane

Motor

Bearing

Outlet connection to rotary pump

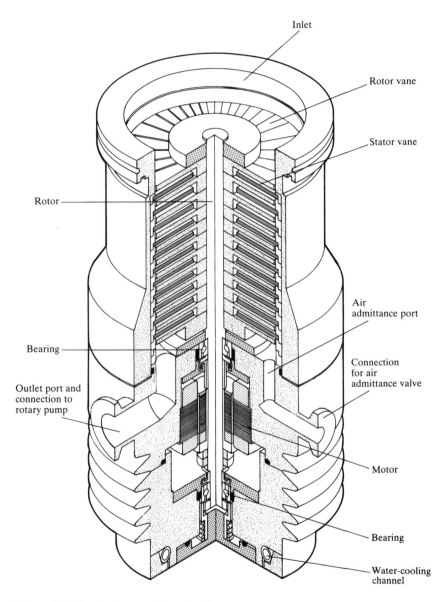

Inlet

Rotor vane

Stator vane

Rotor

Air admittance port

Bearing

Connection for air admittance valve

Outlet port and connection to rotary pump

Motor

Bearing

Water-cooling channel

Figure 8.5   Vertical-type turbomolecular pump

'running-up' time of about one minute is required to bring a small-sized pump up to speed without excessive current overload; large pumps may take several minutes.

For smooth running the rotor needs to be in a very precise state of dynamic balance. Any unbalance that develops will be noticeable due to an increased level of noise and vibration during running. Prolonged running with excessive unbalance will cause damage to the precision ball-bearings, resulting in a further increase in pump noise. Once the bearings are worn, they begin to generate heat at an increasing rate. The bearing tracks rapidly deteriorate and total bearing failure eventually follows. Bearing wear may also be accelerated by contamination of the bearing lubricant with dust or corrosive substances originating from the pumping gas load.

Pumps with magnetically levitated rotors are also available (see Figure 8.6). These pumps are likely to be used in applications that require absolutely hydrocarbon-free pumping, corrosion resistance or extremely low vibration. Examples of their application are in nuclear physics/engineering, surface science experiments, mass spectrometry, electron beam lithography and many others.

The rotor is continuously suspended by magnetic levitation by use of a combination of electromagnetic and permanent-magnet bearings. Supported in this way, without mechanical contact, the rotor has no need for lubrication, thus eliminating any possibility of hydrocarbon contamination. This frictionless condition provides a very low operating temperature so that cooling is not required. They thus have the following advantages over pumps using oils or greases:

- No frictional contact of bearings
- No lubrication
- Free from hydrocarbons
- Low noise and vibration
- No cooling
- Mounted in any orientation

The accompanying control unit contains an internal battery which is available as an emergency back-up. This will keep the rotor supported while it slows down in the event of a mains power failure. In the event of some major failure in the suspension system or if there is a sudden massive air inrush, the rotor 'lands' on a set of back-up 'dry' bearings which safely support the rotor as it comes to rest. After two such 'touch-down' events the dry bearings must be replaced.

These pumps and conventional lubricated types can be used in magnetic and radiation fields, providing the local conditions are known and are within acceptable limits. Excessive magnetic field strengths may result in severe heating of the motor rotor by induced eddy currents. Pumps can always be magnetically screened to some extent if field strengths are a real problem.

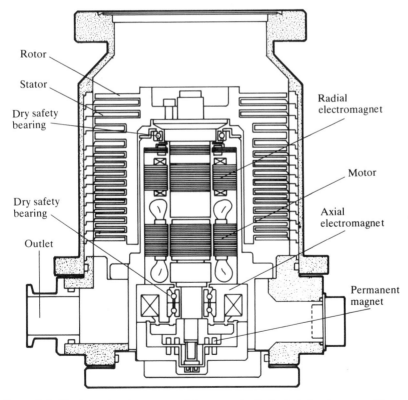

Rotor

Stator

Dry safety
bearing

Radial
electromagnet

Motor

Dry safety
bearing

Axial
electromagnet

Outlet

Permanent
magnet

Figure 8.6 Turbomolecular pump with magnetically levitated rotor assembly

Obviously, ingestion of foreign objects into any turbomolecular pump during operation will cause severe damage to the rotor and stator blading. In order to reduce this risk most pumps are provided with a wire mesh inlet guard.

A bakeout band is sometimes used to heat the inlet area of the pump to accelerate outgassing and enable lower pressures to be achieved. The bakeout band is fitted around the stainless steel envelope of the pump just below the inlet flange. Moderate bakeout to 100 °C can usually be performed on most pumps when operational.

8.3 Operational aspects

Start-up

Pumps are run up to full rotational speed under vacuum. On small-volume systems the turbopump and rotary pump can be started together, provided the system will reach 1 mbar before the turbomolecular pump has reached full rotational speed.

Large systems should be roughed out to about 1 mbar before the turbomolecular pump is started. A foreline trap is an optional accessory that can be used to minimize back-migration of oil vapour from the rotary pump. It is normally recommended where the highest system cleanliness is required. Although when the pump is operating the compression ratio for hydrocarbons is so high that the degree of back-migration is imperceptible. Normally a high vacuum valve is not required, and there is no need for a baffle or liquid nitrogen trap.

Shut-down

Venting of the turbomolecular pump when it is not operating is important for pumps using lubricants. The pump should be vented from the high vacuum side while the rotors are still in motion, thus preventing transport of hydrocarbon contaminants from the backing side to the high vacuum side. If a turbomolecular pump is stopped under vacuum, either by mal-operation or an interruption to the mains electrical supply, lubricant vapour can back-migrate into the high-vacuum system. Thereafter, considerable problems can be experienced in obtaining a satisfactory clean-up of the vacuum system because the mouth and vanes of the turbomolecular pump can only withstand limited baking. Automatic venting with dry nitrogen or air in the event of a power failure is recommended.

A typical operational sequence of a turbomolecular pump system like that shown in Figure 8.7 is:

1. Close the air admittance valve.
2. Start the rotary pump and pump the vacuum system to less than 1 mbar.

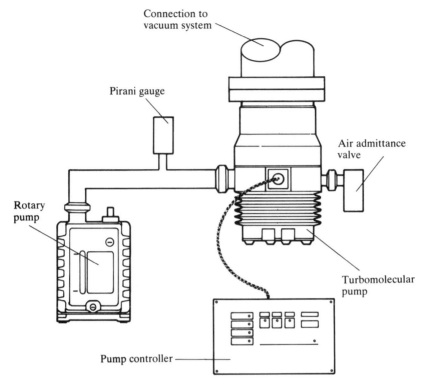

Figure 8.7   Turbomolecular pump system

3. Start the turbomolecular pump and allow working pressures to be reached.
4. Carry out the process.
5. Switch off the turbomolecular pump.
6. Switch off the rotary pump.
7. Admit dry nitrogen/air via the admittance valve while the turbomolecular pump is slowing down.

A typical system capable of operating in this manner would be a $200\,\mathrm{l\,s^{-1}}$ turbomolecular pump backed by a two-stage 5 or $8\,\mathrm{m^3\,h^{-1}}$ rotary pump, pumping a volume of 25 to 30 litres.

8.4 Maintenance of turbomolecular pumps

Routine maintenance is confined to making sure there is sufficient fluid in the oil reservoir of oil-lubricated types or to relubrication of the bearings of grease-lubricated types.

If the rotors and stators become contaminated it may not be possible to achieve the desired ultimate pressure rapidly. It will then be necessary to clean the pump in order to remove the contamination. If it is heavily contaminated then the pump may need to be stripped and cleaned following service instructions. Some manufacturers advise that slight contamination can be removed without dismantling the pump. This is achieved by immersing the pump upside down in a vessel filled with cleaning solvent (in a well-ventilated location observing all necessary precautions) (see Figure 8.8). This should be an organic solvent chosen to suit the contaminant. Suitable solvents may include trichloroethane, ethanol, isopropanol and trichlorotrifluoroethane.

The maximum depth of solvent must not exceed the recommended level to avoid damaging the motor and the bearings. Before undertaking such cleaning check with the manufacturer first.

Possible causes for problems in turbomolecular pumps are given in Table 8.1.

Table 8.1 Troubleshooting turbomolecular pumps

Symptom	Possible causes
1. Turbopump will not start	(a) Fuse blown
	(b) Thermal overload operative
	(c) Fault with motor windings
	(d) Rotor seized
	(e) Faulty controller
2. Turbopump trips into overload during operation	(a) Inlet pressure too high
	(b) Bearing defective
3. Pump very noisy and/or large vibration	(a) Pump rotational speed is the same as the resonant frequency of the attached system
	(b) Rotor out of balance
	(c) Bearing defective
4. Turbopump runs hot	(a) Check cooling water or that air cooling fan is operational
5. Ultimate pressure not reached	(a) Faulty gauges
	(b) Poor conductance between pump and chamber
	(c) High backing pressure
	(d) Pump contaminated
	(e) System contaminated/leaks

8.5 Turbomolecular pump applications

The fact that turbomolecular pumps are able to produce high vacuum, uncontaminated by back-streaming and without the use of traps and baffles, makes them invaluable in certain applications. Examples are in clean surface science studies, gas analysis, certain semiconductor applications and in fusion

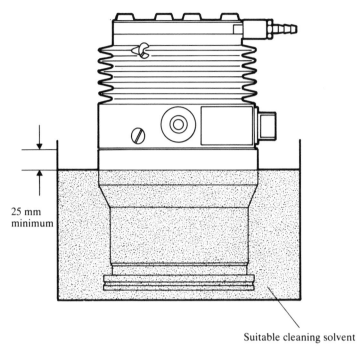

25 mm
minimum

Suitable cleaning solvent

Figure 8.8   Suggested cleaning arrangement for one particular make of turbomolecular pump. (Always check with manufacturer before undertaking cleaning)

experiments. Other features that make these pumps attractive are their rapid operational readiness and relatively maintenance free operation. Standard pumps are not always suitable and certain models have been adapted and developed into models capable of dealing with corrosive or radioactive gases, for example.

Semiconductor applications

Turbomolecular pumps are now widely used in plasma etching applications because of their cleanliness and ability to handle relatively high throughputs of gas continuously. Standard pumps showed poor resistance to corrosion in initial trials and a lifetime of less than 50 hours of pumping chlorinated substances was quoted by one manufacturer. The primary point of attack is on the bearing via the lubricant, leading to failure. Lubricating with PFPE fluids did not significantly improve the situation.

Today's corrosion-resistant turbomolecular pumps have been designed with a gas purge flowing through the bearing and motor region towards the exhaust side of the pump (see Figure 8.9). Gases such as dry nitrogen or argon act as a dynamic barrier against possible corrosive attack by the reactive process gases being pumped.

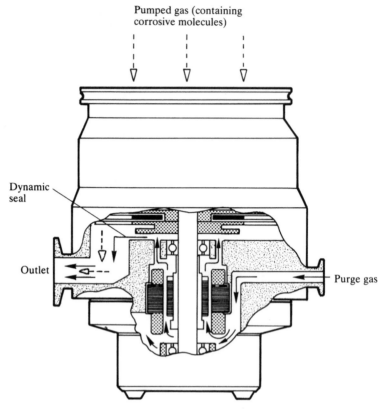

Figure 8.9 Corrosion-resistant turbomolecular pump

Fusion applications

Vacuum pumping of radioactive gases, such as tritium in plasma fusion experiments at ultra-high vacuum, represents a serious safety hazard if the gas is inadvertently released into the environment. Furthermore, tritium can replace hydrogen in organic materials within the vacuum system, making them radioactive and changing their physical properties. It can therefore cause contamination and deterioration of materials such as elastomer seals and pump lubricating oils.

Like hydrogen, tritium has a high rate of diffusion through many materials (including elastomers), making them unacceptable; all joints sealing tritium into the vacuum system must be metal to metal. The pumps have to be operated as a closed system to prevent the relese of tritium into the environment.

Turbomolecular pumps are particularly suitable for such operating conditions and special tritium-resistant designs have been developed. Alternatively, 'dry' non-lubricated turbomolecular pumps with magnetic levitation of the rotor system are available; these pumps are lubricant free and can be readily adapted to tritium applications.

8.6 Summary: points with respect to turbomolecular pumps

1. Turbomolecular pumps have constant pumping speeds between 10^{-3} and 10^{-10} mbar and can achieve UHV pressures without the use of liquid nitrogen traps.
2. The turbomolecular pump compression ration is very gas dependent—low for light gases, e.g. hydrogen, and high for heavy gases, such as hydrocarbon vapours.
3. Turbomolecular pumps are operated in conjunction with a rotary pump.
4. Turbomolecular pumps are let up to atmosphere when not in use. This prevents pump/system contamination by the turbomolecular pump lubricant.
5. On small systems, the turbomolecular pump and rotary pump can be switched on together. On larger systems, the turbomolecular pump should be switched on at 1 mbar.
6. The pump quickly reaches maximum rotational speed; the time taken is generally much less than warm-up time for diffusion pumps.
7. The turbomolecular pump rotational speed is high, i.e. 10 000 to 60 000 rev min^{-1}. Solid objects entering the pump will cause very serious damage!
8. Bearings have a limited life (typically 20 000 hours of continuous running is quoted). Changing may be difficult. Accurate balancing is necessary so expensive specialized equipment may be required. However, with most modern pumps it is possible to change the bearings without the need for rebalancing.
9. With water-cooled turbomolecular pumps, the water should only flow when the pump is running. If the water cooling is left on at atmospheric pressure it may be possible for moisture to condense within the pump.
10. The power consumption of a turbomolecular pump is less than a diffusion pump.

9

Cryopumps

9.1 How cryopumps work

Cryogenic pumping is the process of 'freezing' gas or vapour out of a vacuum system onto very cold surfaces. In principle, any gas or vapour can be pumped provided that the surface temperature is low enough so that condensed gas or vapour will remain on the surface. Cryogenic pumping is a clean form of pumping compared with, for example, a diffusion pump which contains oil as a working fluid. Cryopumping using a cryopump does not present any potential system oil-contamination problem if the pump should fail, since the pump does not contain any oil. In addition, the pumped gas load is only released when the pump is allowed to warm up.

The modern cryopump is a closed-loop refrigerator using gaseous helium to produce temperatures down to 10 K ($-263\,°C$) (see Temperature scales—Appendix F).

The flow of helium is cyclic and is simply illustrated in the cooling circuit shown in Figure 9.1a. The main components are an expander piston made from an insulating material contained in a cylinder and a source of high-pressure (HP) helium gas supplied by a compressor to the cylinder when valve A is open. This gas is typically at 20 bar and ambient temperature. Valve B is in the exhaust line leading to the low-pressure (LP) side of the compressor. With the piston at the top end of the cylinder (the nearest end to the compressor) and with valve B closed and valve A open, the piston is forced downward and the top of the cylinder fills with compressed gas. When valve A is closed and valve B is opened, the gas expands into the low-pressure discharge line and cools. The resulting temperature gradient across the cylinder wall causes heat to flow from the load into the cylinder, warming the helium to a temperature somewhat below that at which it entered the cylinder. The piston is then driven up to force the remaining cooled, expanded gas back to the compressor.

This elementary system, while workable, would not produce the extremely low temperatures required. An improvement can be obtained if the incoming high-pressure gas is cooled indirectly by the exhaust gas before it reaches the cylinder. This is achieved by a regenerator (shown in Figure 9.1b), which extracts heat from the incoming gas, stores it and then releases it to the exhaust stream. The regenerator is tightly packed with a metal of high heat

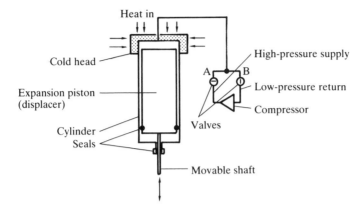

(a) Elementary cooling circuit

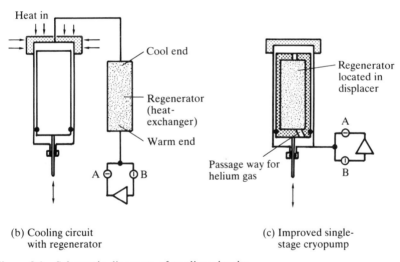

(b) Cooling circuit
with regenerator

(c) Improved single-
stage cryopump

Figure 9.1   Schematic diagrams of cooling circuits

capacity arranged so that the surface-area-to-volume ratio is large. Small
spheres, wire mesh or fine gauze are used. Lead shot and copper or phosphor
bronze gauze are the most commonly used packing materials. Even though
the regenerator is tightly packed, there is not much flow resistance.

During normal operating conditions the regenerator will have a temperature gradient. Ambient temperature helium entering from the warm end will
give up heat to the metal in the regenerator; cold gas entering in the reverse
direction from the cooler end will absorb heat from the metal. The
regenerator can transfer thermal energy from the incoming to the outgoing

helium quickly and with great efficiency. A single-stage machine of this design can achieve temperatures in the 30 to 60 K range.

It is possible to incorporate the regenerator inside the displacer (see Figure 9.1c), thus reducing the cryogenerator to a single cylinder. The high- and low-pressure valves are generally built into a single unit and the displacer can either be driven up and down mechanically or, with some extra refinements to the gas supply channels, it can be moved pneumatically by the working gas itself. The latter approach is of particular elegance because of the consequent reduction in moving parts, and Figure 9.2 shows a schematic section of a two-stage cryopump incorporating this feature. Here, the valve disc/motor assembly comprises a drive motor and rotary valve disc mechanism which functions to control the flow of helium gas into the expansion volumes of the first and second stages. Lower temperatures can be achieved using two-stage machines. The first stage operates in the region of 40 to 70 K, whereas the second stage operates in the region of 10 to 20 K. The exact temperatures depend on the heat load and capacity of each stage.

A complete cryopump unit consists of three parts (see Figure 9.3): a gaseous helium compressor, high- and low-pressure gas lines and a combined two-stage cold head, baffle and pump body assembly.

The compressor unit consists of a standard, water-cooled, oil-lubricated compressor, with a means of oil separation to prevent carryover of oil mist and vapours in the compressed helium gas. Any oil circulating with the helium would solidify at the cryogenic temperature in the cold head and prevent the cryopump from functioning. The heat produced in compressing the helium is removed by a water-cooled heat-exchanger in the compressor unit. Each stage of the cold head is connected to an extended surface (cryopanel) onto which the gas can freeze (see Figure 9.4). The second-stage cryopanel is completely surrounded by the combined first-stage cryopanel and louvre array. This thus acts as a radiation shield for the second stage. In addition to heat shielding, the panel also acts as a very high speed pumping surface for system vapours such as water and relieves the second-stage cryosurface from this kind of load. The second-stage cryopanel is used to freeze out gases such as nitrogen, oxygen and argon which pass through the louvre. The inside of the second stage is coated with charcoal, which is used to adsorb gases that will not freeze at the second-stage temperature. These are hydrogen, helium and neon.

A hydrogen vapour pressure thermometer gauge or an encapsulated diode thermometer attached to the second stage is used to monitor its temperature. In general, the extremities of the cryopanel will be at a slightly higher temperature than may be indicated on the thermometer, and if there is a layer of cryodeposit already laid down the surface temperature will certainly be warmer than that measured. The thermometer temperature is not, therefore, a reliable guide to ultimate pressure, although it is nonetheless a good indication as to the total heat load on the second stage of the cryogenerator

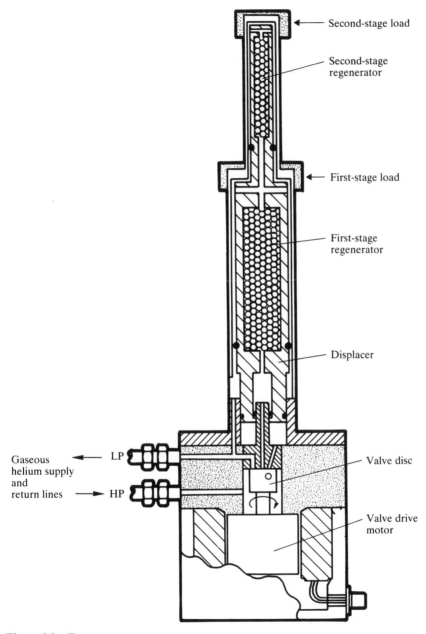

Figure 9.2 Two-stage cryopump

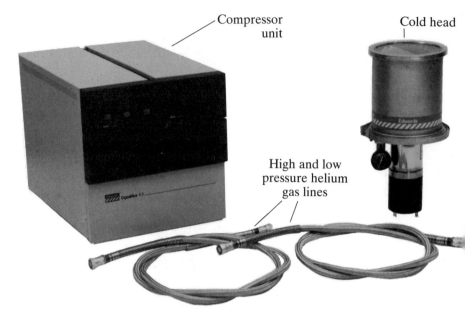

Figure 9.3 A modern cryopump cold head and compressor unit

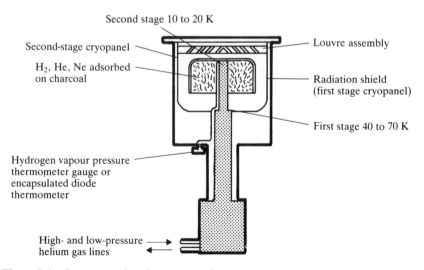

Figure 9.4 Cryopump showing cryopanel arrangement

and can be used as such to determine when the cryopanels are heavily loaded with deposit and regeneration is required.

9.2 Cryopump system layout

Cryopumps take a long time to cool down and warm up again, e.g. typically 1 to 3 hours for cool-down. Most cryopumps are installed equipped with a high vacuum isolation valve (shown as valve V_1 in Figure 9.5), which allows the cryopump to be kept continuously operating under vacuum, in situations where the chamber must be periodically exposed to atmospheric pressure. The use of a valve thus saves cooling and warming time. The valve can also be used in conjunction with a pressure set-point controller, continuously receiving system pressure data from a pressure gauge to provide automatic closure in the event of a sudden rise in pressure due to a gross leak. The cryopump does not need a backing pump, but must be evacuated to a low pressure before it is cooled down. A mechanical two-stage rotary pump is usually used for this purpose. In order to avoid any oil back-migration from the rotary pump, it is normal to fit a foreline trap in the pumping line. This can if necessary be bypassed during high throughput roughing. Valve V_4 is an optional valve that allows the rotary pump to be removed/switched off or otherwise isolated from the cryopump, after initial evacuation is completed. Valve V_3 is used to isolate the gauge head from the pump; the reason for

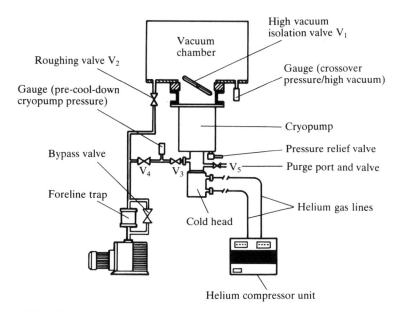

Figure 9.5 Cryopump system layout

this is discussed in Section 9.6 (item 4). The cryopump base flange is fitted with an integral pressure relief valve to release gases evolved during pump regeneration. These gases can be purged out of the pump by a purge gas supplied via valve V_5.

9.3 Operation of a cryopump system

Initial evacuation and cooling (see top left-hand diagram of Figure 9.6)

Before starting the cool-down of a cryopump it is advantageous to seal it off from the main chamber using the high vacuum isolation valve V_1. There is a tendency for the activated charcoal on the second-stage cryopanel to absorb gases, particularly water vapour, during cool-down, long before the main condensing surfaces are cold enough to function fully. Clearly, the less gas that is available to be adsorbed during the cool-down period the greater will be the available storage capacity of the charcoal during subsequent cryopumping. In the arrangement shown opposite the correct start-up procedure is, therefore, to ensure that the roughing valve V_2 and the main high vacuum isolation valve are closed and valve V_3 is open. The cryopump can then be roughed out in isolation prior to commencing cool-down. This surrounds the cryopanels with an insulating vacuum, minimizing heat transfer between them and the pump body, and thus greatly assisting with cool-down. The cryopump is switched on when the pressure reaches typically 5×10^{-2} mbar. The rotary pump continues to 'assist' the cryopump until the cryopump is cold enough to start pumping and produce pressures lower than the rotary pump (typically this takes about 10 minutes for a 200 mm inlet diameter cryopump). At this point, valve V_3 is closed and the cryopump then continues to operate unbacked.

Roughing out the chamber to the 'crossover pressure' (pump-down) (see top right-hand diagram of Figure 9.6)

The roughing valve V_2 may be opened to allow the rotary pump to evacuate the main chamber. The pressure is then reduced to what is called the 'crossover pressure', i.e. the highest chamber pressure from which the cryopump will safely evacuate when fully operational. The pressure limitation is related to the maximum thermal load that the cryopanels can handle. In particular the term 'cryopump capacity' is a measure of the mass of gas that can be impulsively handled by the pump without the second-stage panels warming above 20 K. A value of 'crossover capacity' is normally supplied by the manufacturer and is generally expressed in millibar-litres.

Crossover capacity = chamber volume × maximum cryopump evacuation pressure (crossover pressure)

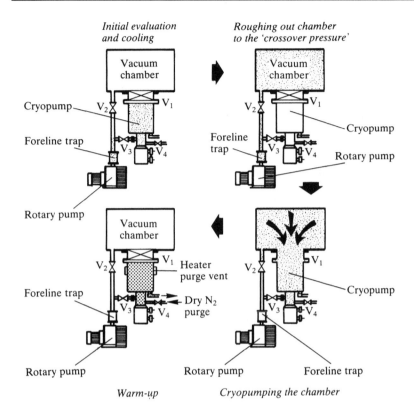

	Valve condition				Time required
	V_1	V_2	V_3	V_4	
Initial pressure and cooling	Closed	Closed	Open	Closed	1–3 hours
Roughing	Closed	Open	Closed	Closed	Less than 1 hour
Cryopumping	Open	Closed	Closed	Closed	Dependent on system
Regeneration	Closed	Closed	Closed	Open	Typically 2–6 hours

Figure 9.6 The various stages in the operation of a cryopump system (start at top left-hand diagram)

Thus, if crossover capacity is given as 120 mbar-litres, for a chamber of 100 litre volume, the cryopump should be brought in only when the pressure has fallen below $\frac{120}{100}$ mbar, i.e. 1.2 mbar.

Bringing in the cryopump at too high a pressure will raise the second-stage panel to a temperature above 20 K, with the possible release of previously pumped hydrogen or helium from the sorption surfaces. On the other hand, if the cryopump is brought in at too low a pressure (i.e. below 1 mbar) this will result in a long roughing cycle and possible back-migration of rotary pump oil vapour to the vacuum chamber. Note that even with a foreline trap fitted there may be a small amount of back-migration (1 per cent) which may be a significant contaminant with repeated roughing of the chamber.

Cryopumping the chamber (see bottom right-hand diagram of Figure 9.6)

When the cryopump is fully cooled and the 'crossover pressure' has been reached, the pump chamber is isolated from the rotary pump (V_2 closed) and the high vacuum valve V_1 is opened. At this point the rotary pump could be shut down if required. In most production applications, the cryopump will typically operate for several weeks before having to be warmed up to be regenerated.

Regeneration (see bottom left-hand diagram of Figure 9.6)

Regeneration is the process whereby the trapped gases are removed from the cryopump and it is returned to its full rated pumping speed and capacity. This is done by closing the high vacuum valve V_1, turning off the compressor and allowing the pump to warm up. As the pump warms up, the pressure within the cryopump housing will increase and will rise above atmospheric pressure. At a pressure of a few tenths of a bar above atmospheric pressure the gases will be released through a spring-loaded relief valve. The warm-up process may be accelerated by introducing a flow of dry nitrogen gas and by heating the pump casing. Nitrogen gas not only facilitates warm-up but it will also dilute potentially explosive gas mixtures that may be released during warm-up. It will further dilute chemically reactive or poisonous gases. Dangerous materials should be vented in a safe fashion.

Pumped water vapour can remain as a liquid when released during warm-up and will need to be pumped away during regeneration by the rotary pump. Once initiated, it is important that regeneration is continued to completion and all released gases (or liquids) removed from the pump. If just switched off and left, particular aggressive compounds that may have been pumped could damage the cryopanels if they remain in the pump for long periods of time.

Operation time between regenerations

The need for regeneration is evident when the cryopump is not able to attain the proper vacuum levels, although the temperature of the second stage is less than 20 K. The frequency of regeneration will depend on the pump's capacity for a particular gas species and on the throughput of that gas. Some examples are:

1. Sputtering using argon at a pumping speed of $1000 \, l \, s^{-1}$ and a pressure of

3×10^{-3} mbar will give an operational lifetime before regeneration of about 93 hours of continuous running.

2. Pumping hydrogen (as a carrier gas in an ion implanter) at 10^{-6} mbar with a hydrogen pumping speed of $2000 \, l \, s^{-1}$ would require regeneration of the pump about every two months.

9.4 Applications

An important difference between the pumping performances of the diffusion pump and the cryopump is that the cryopump can pump vapours much faster (about three times as fast for water vapour). Unbaked systems, where the remaining gas load may be mainly water vapour, can therefore be pumped down more quickly and to somewhat lower pressures by the cryopump. More importantly, during a process where much water vapour is released (such as a coating process) the cryopump is better able to 'hold' a low pressure during processing.

With the exceptions mentioned below, there are few applications for which the cryopump cannot be considered.

The cryopump has a very limited capacity for the light gases such as hydrogen (which can be produced during some processes), helium or neon. This may be the limiting factor in deciding the frequency of regeneration and may make them clearly unsuitable for pumping duties that require a high throughput of these gases.

Cryopumps are not recommended for long-term operation in high-intensity radiation areas since some of the internal organic parts of the cold head may be degraded.

For pumping corrosive gases and vapours the cryopump collects and concentrates corrosive effluents on its cold surfaces. These do no damage while they remain cold, but when the cold surfaces are warmed for regeneration, strong solutions of corrosives may run on these surfaces and drip onto other parts of the pump and damage may result. Cryopumps may similarly concentrate toxic substances.

Cryopumps are not tolerant of dirty and dusty loads and their use for such duties (such as on vacuum furnaces) should be avoided. The cryopump has a large capacity for argon, and can maintain high pumping speed up to quite high pressures. Its use on sputtering applications has been quite widespread. The impossibility of accidental oil contamination (which might occur with diffusion pumps) of the chamber appeals to some users.

Cryopumps must not be used in applications where the cryopanels will be exposed to excessive radiant heat loads from the surroundings, without cooled optical baffles.

For pumping clean systems down to a very low pressure for prolonged periods, metal gasket versions are available, and in this case the pump can be mildly heated when the system is baked. Gas loads are small and the pump can operate for very long periods without regeneration. In mildly baked systems the outgassing load will contain some water vapour and the high vapour pumping speed of the cryopump is a positive advantage. Pressures in the 10^{-10} mbar range are possible.

Cryopumps for sputtering systems

Sputter coating is widely used as a technique for thin-film deposition. In such processes argon may be present at pressures above 10^{-3} mbar. A conventional cryopump under such operating conditions would be exposed to high gas loads and unacceptably high thermal loads. This would result in the warming of the cryopanel and release of previously condensed gases.

Throttling the argon throughput into the cryopump by means of a conventional throttle valve results in a reduced pressure in the pump while maintaining the pressure in the chamber. However, it has the disadvantage that the pumping speed for all other undesirable residual gases is also

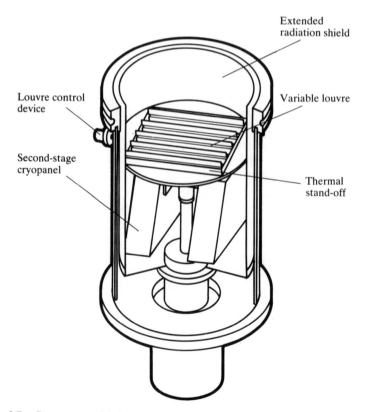

Figure 9.7 Cryopump with integrated throttle

reduced, including water vapour whose presence adversely affects the quality of deposited layers.

A development which solved this problem is a cryopump (shown in Figure 9.7) in which the throttle valve is integrated in the pump and the radiation shield pumping surface is extended above the valve. The valve can be considered as a variable-pitch louvre baffle whose conductance can be adjusted externally. It is mounted from and thermally connected to the pump body. This maintains the valve at or near ambient temperature and therefore prevents buildup of deposits on the louvre assembly, which would otherwise impede the mechanical operation of the valve and gradually reduce its conductance. Although closing this valve reduces the argon throughput, as required, a high pumping speed for water vapour is still maintained by the extended pumping surface, while the overall thermal load on the pump is reduced.

Another solution is to use a variable-aperture module which attaches to the first-stage cryopanel in place of the conventional louvre assembly. It thus operates at 40 to 70 K and acts as a throttle for the process gas, but maintains a high pumping speed for water vapour. The variable-aperture mechanism (rotor) is air actuated to the fully open position during the initial system pump-down. During the process, the rotor can be actuated closed to any preset position to control the process gas pumping speed and hence system pressure. Special Teflon insulators isolate the cold operating surface from the room temperature actuator mechanism. The stator–rotor spacing and positive actuating action eliminate the possibility of the mechanism freezing during normal operation.

9.5 Looking after cryopumps

Maintenance periods depend very much on the manufacturer and model. It is recommended that an operating log for recording system parameters is kept while operating a high vacuum pumping system. Time, temperature and pressure readings during operation under both normal load and cool-down should be recorded. The comparison of recordings, logged over a period with correct operating parameters, can provide an early indication of a drop in performance and can also be useful in the diagnosis of faults.

The cryopump compressor and cold head

Compressors can run almost indefinitely, the only routine maintenance task necessary to maintain efficient operation being the renewal of the compressor unit adsorber every 6000 to 9000 hours (typically this can take about $\frac{1}{4}$ hour to accomplish).

Figure 9.8 shows a block diagram of a typical cryopump compressor layout. The compressor module includes a water-cooled, hermetically sealed, oil-lubricated, single-cylinder compressor. The oil separator is a unit for removing the majority of lubricating oil from the high-pressure helium stream leaving the compressor. The separated oil is returned to the compressor via a capillary connected to the low-pressure side of the compressor. The helium leaving the separator still contains small amounts of oil vapour, and this is passed onto the adsorber unit which contains activated charcoal. The activated charcoal removes the oil but eventually becomes saturated and will need replacing. The adsorber unit of most compressors can be replaced as a whole without releasing the pressure in the system, having self-sealing helium supply and return lines. Removal and replacement of the adsorber unit is therefore relatively easy. Failure to replace the adsorber at the specified times will eventually result in oil vapour freezing in the cold head and subsequent prevention of movement of the displacer. Spent adsorbers are normally depressurized before disposal for safety reasons.

Failure to replace the adsorber regularly will result in cryopump malfunction.

It should not be necessary to add helium gas to the compressor unit during normal operation. If it becomes necessary to add helium frequently, i.e. every

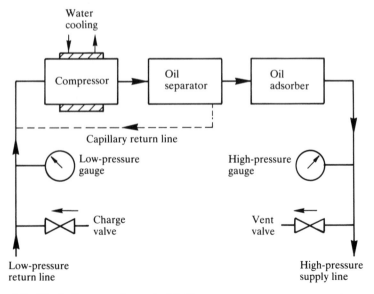

Figure 9.8 Block diagram of a typical helium cryopump compressor unit

few months, the installation should be checked for minor leaks. Leaks are generally caused by incorrectly tightened self-sealing couplings or badly seated relief valves.

Charging may be carried out with the compressor either running or shut down, following the manufacturer's recommended procedures. If charging of the cryopump is undertaken, the use of helium gas of sufficient purity is of the utmost importance. Helium gas that is 99.995 per cent pure should always be used. Helium gas generally available for industrial use, e.g. in helium leak detection and welding, is not usually sufficiently pure and must *not* be used.

Contamination of the helium gas circuit of a cryopump is usually indicated by sluggish or intermittent operation of the cold head displacer piston. In severe cases of contamination the displacer piston will seize.

The cold head expander module seals will eventually need replacement. Life of the seals varies, figures of around 10 000 hours being claimed. Normally replacement can be achieved in under 2 hours.

Possible causes for problems in cryopumps are given in Table 9.1.

Table 9.1   Troubleshooting a cryopump

Symptom	Possible causes
1. Vacuum system has high base pressure with a cryopump temperature less than 20 K	(a) Leak in vacuum system or cryopump (b) Second-stage cryopanel has reached its capacity limit (c) Cryopanel is loose and is not making good thermal contact
2. Vacuum system has high pressure with a cryopump temperature above 20 K	(a) Second-stage cryopanel has reached its capacity limit (b) Excessive thermal load on the cryopump from the vacuum system (c) Cold head faulty
3. The cryopump does not cool down to normal operating temperature or takes an excessively long time	(a) Compressor has low helium gas charge (b) Leak in vacuum system (c) High partial pressure on non-condensables (d) Helium gas contaminated (e) Insufficient degree of pre-evacuation
4. Compressor fails to start or runs briefly and then shuts down	(a) Insufficient gas charge (b) Fuse blown; circuit breakers are tripping (c) Overload protecter actuated (d) Temperature protector actuated
5. Cold head motor is not operational (compressor running)	(a) Motor fuse is blown (b) Faulty connector or wiring (c) Motor is defective

9.6 Cryopump safety

From a safety standpoint, it must always be remembered that the cryopump is a capture pump. Gases accumulated during operation will be released when the pump is warmed up. These gases may be explosive, corrosive or toxic depending upon the application. During operation the elements are so cold that chemical reactions are virtually stopped. It is only during the time they are close to room temperature that they can become a problem. The following safety features are recommended:

1. Where a high vacuum valve is used above the cryopump it should be arranged to close automatically in the event of refrigerator shut-down. Otherwise the vacuum chamber must be designed to withstand the overpressure of pumped gases as set by the relief valve.
2. The pressure relief valve must not be blocked, modified, removed or altered in any way.
3. Toxic or combustible gases should be ducted to a safe area after release through the relief valve.
4. A hazard exists when regenerating flammable mixtures of pumped gases where an ignition source is present. For example, hydrogen–air mixtures are flammable over a wide range of concentrations. An ionization gauge or any other type of possible ignition source should not be located below the high vacuum valve in direct connection with the cryopump.
5. Both cryopanel arrays will remain cold for some time after the compressor is turned off and cannot be handled without the danger of frostbite. The pump should not be opened until the panels are at room temperature.

9.7 Glossary

Cryopumping using a closed-loop heat-transfer mechanism is a comparatively new way of producing vacuum, when compared, for example, with the long-established diffusion pump. It may therefore be useful to include here a glossary of some of the terms commonly used. A more general glossary of vacuum terminology will be found in Appendix B.

Capacity. The amount of a particular gas that can be pumped before the ultimate pressure begins to deteriorate. The value will be different for various gases and vapours and is usually quoted in standard (atmospheric) litres at 10^{-6} mbar.

Cold head (also known as the expander). Houses the displacer mechanism attached to the cryopump that uses high-pressure helium gas to generate useful cooling power by expansion.

Condensable gases. Gases that can be condensed in vacuum at temperatures that can be achieved economically.

Cooling power. Refrigerating capability in watts stated at specific temperatures of the first and second stages. High powers giving fast initial cooldowns, the capability to pump gas continuously at high rates and protection from overheating due to high radiative heat loads onto the cryopanels.

Crossover capacity. A measure of the impulsive mass of gas the cryopump can handle and condense on the second-stage cryopanel without warming to a temperature above 20 K—usually quoted in millibar-litres.

First-stage cryopanel (radiation shield). The nickel-plated copper cryopanel traps water vapour on its surface and shields the second-stage cryopanel from radiant heat originating from the cryopump body.

Hydrogen vapour bulb. A means of measuring the temperature of the second stage of the cold head. Utilizes a capsule-type pressure gauge to measure the vapour pressure of a small pool of liquid hydrogen trapped in a small bulb mounted on the second stage of the cold head. The bulb is connected to the pressure gauge via capillary tube; the gauge is calibrated in kelvin.

Louvre. The louvre is a series of nickel-plated baffle plates attached to the first-stage cryopanel. It traps water vapour and shields the second-stage panel from radiant heat originating from the vacuum system.

Non-condensable gases. Hydrogen, helium and neon can only be condensed in vacuum on a surface at an extremely low temperature that would be costly to achieve in practice. As such they are usually adsorbed into a bed of cooled sorbent. The term non-condensable gas is therefore not strictly true but suffices to differentiate between these gases and those more easily condensed.

Regeneration. Consists of essentially warming up the pump to release pumped gases, re-evacuating the pump and cooling down again to re-establish full pumping speeds and capacity.

Second-stage cryopanel. The second-stage cryopanel fits inside the first-stage panel. On the inside of the second-stage panel, the charcoal adsorbs helium, hydrogen and neon. Nitrogen, oxygen and argon freeze onto its outside surface.

Silicon diode. Another means of measuring second-stage temperature using the variation in electrical resistance with temperature of a silicon diode mounted on the second stage. Uses characteristics of a silicon diode and its variation with temperature to give a measure of temperature.

9.8 Summary: points with respect to cryopumps

1. Accidental organic contamination of the system by the pump is impossible. It is ideal where absolute cleanliness is required.

2. Ultimate pressures of 10^{-10} mbar are possible.
3. The pumping head is compact compared with a diffusion pump. It can be mounted in any orientation.
4. No liquid nitrogen is required.
5. Periodic regeneration is necessary. For 'normal' duties the pump has sufficient capacity to be regenerated at convenient non-operational periods, e.g. weekends.
6. A rotary pump is necessary for initial evacuation of the cryopump and chamber and for regeneration, but the pump is not needed for cryopumping itself.
7. The air pumping speed is comparable to equivalent sizes of diffusion pump. The cryopump has greatly superior speeds for water vapour (except where a diffusion pump has a liquid nitrogen trap).
8. Pressure relief valves are necessary to prevent dangerous internal pressure buildup during regeneration.
9. Cryopumps can handle dangerous gases (but care is needed during regeneration).

10

Sorption and getter pumps

10.1 Sorption pump

The sorption pump is an alternative to the oil-sealed rotary pump as a roughing pump and is used where no organic contamination whatsoever can be tolerated. In most cases the sorption pump has an aluminium or stainless steel body with internally extruded or fabricated heat-transfer fins (e.g. see Figure 10.1). The pump is filled with a very porous material known as molecular sieve. Typically this is aluminium calcium silicate in the form of pellets 5 to 8 mm long and 3 mm diameter. When this is cooled to liquid nitrogen temperatures gases are trapped on its extensive surfaces. The molecular sieve is made up of cavities connected by minute channels or

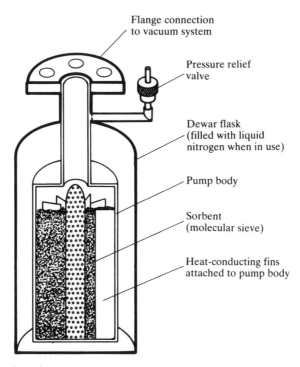

Flange connection
to vacuum system

Pressure relief
valve

Dewar flask
(filled with liquid
nitrogen when in use)

Pump body

Sorbent
(molecular sieve)

Heat-conducting fins
attached to pump body

Figure 10.1 Sorption pump

interconnecting pores. Those gases that do not readily condense at liquid nitrogen temperatures enter the pores and travel into the bulk of the material. Cooling is achieved by immersing the pump body in a dewar of liquid nitrogen. Collision of the gas with the cooled material causes progressive loss of thermal energy and the gas thus finally becomes trapped (sorbed). Upon completion of the pumping cycle, the pump is allowed to warm up to ambient temperature by removal of the liquid nitrogen; the sorbed gas desorbes from the sieve and is expelled to atmosphere through an automatic vent valve which is integral with the pump body.

The recommended molecular sieve pore diameter is about 0.4 nanometres (nm) compared with molecular diameters of 0.32 nm for nitrogen and 0.29 nm for oxygen. The internal surface-area-to-mass ratio is typically $550 \, m^2 \, g^{-1}$; 1 kg therefore represents a surface area equivalent to approximately $0.6 \, km^2$ ($0.2 \, mile^2$). Sorption pumps are sized according to the weight of molecular sieve they contain, typically ranging from 300 g to 1.2 kg.

The design and performance of the pump are mainly dictated by three factors:

1. *The saturation effect.* Water vapour is strongly sorbed by sorbent materials and unfortunately is not released when the pumps are warmed to room temperature to release the sorbed gases. The accumulation of this water vapour results in a deterioration of pumping performance which can be restored by regenerating the sorbent by heating typically at 300 °C for two hours. Pumps should not be left open to atmosphere when not in use.
2. *The preferential pumping of the various gases.* Since the adsorption effect is selective, sorption pumping speeds vary with the nature of the gas. Nitrogen, oxygen, carbon dioxide, water vapour and hydrocarbon molecules are efficiently pumped. Light gases, notably hydrogen, helium and neon are never truly trapped by either condensation or sorption at liquid nitrogen temperature and can begin to back-diffuse into the chamber.
3. *The rate of removal of heat from the pellets.* The design of the pump is a compromise between ensuring that the sieve pellets are adequately cooled and that the pump has sufficient conductance for the gas to reach the sieve material.

Because of the many variables a normal pumping speed curve is not of much use and performance is usually quoted in terms of pump-down curves of chamber volumes. The size of pump required is determined by the system volume, the ultimate pressure required, the nature of the gas being pumped and the number of pumps used. For example, for an ultimate pressure of 4×10^{-2} mbar in a 15 litre system, typically one 300 g pump would be used.

Ultimate pressures of 10^{-2} mbar are possible with a single pump, and 10^{-4} mbar with sequential pumping using two or more pumps. The use of a non-lubricated carbon-bladed rotary-vane pump to rough the system to about 100 mbar is useful since it takes some of the 'load' off the sorption pumps. Saturation of the sorption pump does not occur so readily and regeneration is required less frequently.

For single sorption pump use, the pump is prechilled for about 20 minutes, isolated from the chamber. After the sorption pump has cooled and the system has been roughed out with the carbon-bladed pump, the isolation valve to the rest of the system is opened and the sudden inrush of air carries all gas types into the pump. However, some back-diffusion of difficult-to-pump gases will occur. Single-stage pumps can operate from atmospheric pressure down to about 10^{-2} mbar. For improved ultimate pressure, when two pumps are used, the first pump is valved off soon after the initial inrush in order to limit back-diffusion. The second pump then takes over to continue pumping.

In summary, as there are no moving parts within the pump it is free from noise and vibration and also creates an absolutely hydrocarbon-free vacuum. It is thus ideally suited as a roughing pump for ultra-high vacuum systems that utilize sputter-ion pumps. In comparison to rotary pumps, which can operate continuously, sorption pumps can only be used on a cyclic basis, where gases are frequently released after each pump-down.

10.2 Getter pumps

Cryopumps and sorption pumps depend on the physical adsorption of gases on refrigerated surfaces for the pumping mechanism. This section deals with pumps that rely on chemical sorption processes for trapping the gas. Chemisorption as discussed briefly in Section 2.4 is characterized by a much stronger bond between the gas and surface molecules. The chemisorption of gas molecules on a surface is often referred to as gettering. Getters are used in television tubes to improve the vacuum after sealing from the pump and to maintain the vacuum during the tube's life. Getters are chemically active materials that form stable compounds with gases and thus remove them from the vacuum system. The getter material most commonly used today as a pump in its own right is titanium and the clean active metal surface can be provided in many ways.

Sublimation pumps

Titanium sublimation pumps are the most commonly employed auxiliary pumps for UHV systems. The sublimation pump relies on the sublimation of a titanium source onto a condenser surface. Typically this is achieved by resistance-heating titanium filaments. The deposited active film of the metal is then available to getter gas molecules. The design of the pump is very simple and can consist merely of a flange with electrical leadthroughs and supports for the filaments. This constitutes a cartridge which can be mounted directly into the vacuum chamber (suitably shielded to prevent titanium depositing on feedthroughs, gauges, etc.). Titanium will evaporate onto the chamber walls or onto a suitable baffle mounted around the filament (see Figure 10.2). Better performance is obtained when the titanium is deposited onto a cooled surface. Substrate cooling can be achieved by forced air cooling or water cooling the outside of the walls where the deposit is formed. Alternatively, liquid nitrogen cooled surfaces within the system can be used.

Titanium pumping is selective; it captures getterable gases such as nitrogen, oxygen, hydrogen, carbon dioxide, water vapour and carbon monoxide, forming solid low vapour pressure compounds such as titanium hydride, titanium oxide, titanium nitride, etc. Other types of pumps have to be used to remove helium, argon, etc. At high pressures (above 10^{-5} mbar) the titanium will be consumed as fast as it is being deposited. The pumping speed in this condition is thus dependent upon the amount of titanium that can be laid down. At lower pressures (less than 10^{-6} mbar) it is possible to deposit titanium faster than it is being consumed. The pumping speed is thus

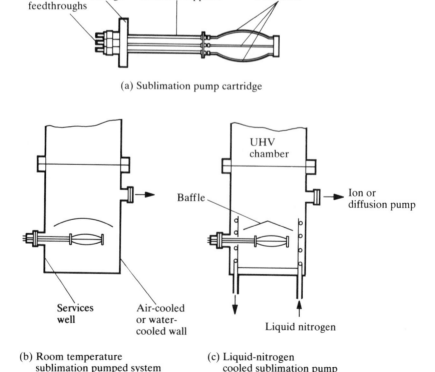

(a) Sublimation pump cartridge

(b) Room temperature sublimation pumped system

(c) Liquid-nitrogen cooled sublimation pump

Figure 10.2 Sublimation pump and pumping arrangements

Table 10.1 Approximate pumping speeds of a fresh titanium film

	Pumping speed $(l\,s^{-1})$ for 1 cm^2 of film area							
Gas	H_2	N_2	O_2	CO	CO_2	H_2O	Ar	He
Surface temp. 20 °C	3	4.5	1·5	9	7.5	3	0	0
-195 °C	10	10	6·2	10.5	9	14	0	0

dependent primarily on the amount of gas that can get into the chamber, i.e. it is limited by the conductance of the system. Intermittent operation is used to extend the life of the filament. In the 10^{-7} mbar range and lower, titanium need only be deposited periodically, with the period becoming longer as the pressure decreases. Table 10.1 gives the maximum pumping speeds in litres per second, ignoring any conductance limitation, for a fresh titanium film of 1 cm^2 area for various gases and surface temperatures.

Titanium sublimation pumps can be used:

1. To produce very high pumping speeds at periods of high system outgassing, e.g. during the degassing of an evaporant prior to deposition.
2. To reduce the ultimate pressure attainable by an order of magnitude. With a clean empty system an ultimate pressure of about 5×10^{-10} mbar is typical for an efficiently designed ion pumped system. Better than 5×10^{-11} mbar would be expected using a titanium sublimation pump as well.
3. To reduce the start-up time of a sputter-ion pump. Operation of the sublimation pump during the start-up phase of an ion pump will help to quickly reduce the system pressure. While this technique is very useful in some cases, regular fast starting has been found to have an adverse effect on the low-pressure performance of the system if pressures below 10^{-9} mbar are required. The ion pump and system are deprived of some of the beneficial effects of operation for a period at high pressure.

Sputter-ion pumps

This pump makes use of ions to sputter a getter film. Cathode material such as titanium is sputtered (i.e. knocked off) by bombardment with high-velocity ions. The deposited titanium film combines readily with active gases in the same manner as the sublimation pump pumps active gases. At the same time, inert gases are removed by ionization and burial on the cathode by sputtered titanium and light gases such as hydrogen and helium by ionization and diffusion into the cathode. This pump therefore removes gases that the sublimation pump cannot.

A typical pump consists of two flat rectangular cathodes with a stainless steel anode between them made up of a large number of open-ended tubes (see Figure 10.3). This assembly, mounted inside the pump body, is surrounded by a permanent magnet. The anode is operated at a potential of about 7 kV whereas the cathodes are at ground potential. This layout is referred to as a diode structure pump. The initiation of the cold discharge depends on the fortuitous production of an electron, e.g. by cosmic rays or by emission of electrons from the cathodes. These electrons travel towards the anode and execute a rotary motion around the magnetic field lines and oscillate to and fro in the axial direction between the cathodes before being collected by the anode (see the enlarged section of a basic cell in Figure 10.3). The long paths of the electrons result in a greater probability that ionizing collisions with gas molecules will occur than if the electrons were able to go directly to the anode. The positive gas ions formed by ionization bombard the cathodes so that the pumping process described above results. At the

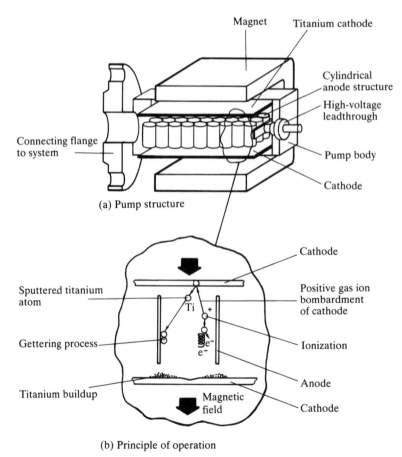

(a) Pump structure

(b) Principle of operation

Figure 10.3 Sputter-ion pump

higher pressures, where much ionization takes place, more current is drawn. At low pressures, less current is drawn. This current characteristic can be used to measure approximately the pressure within the pump.

Pumps with pumping speeds up to $1000 \, l \, s^{-1}$ are readily available, with an operating pressure of 10^{-2} to below 10^{-11} mbar. The full speed of the pump is developed in the pressure range from about 10^{-6} to 10^{-8} mbar. Larger pumps up to $10\,000 \, l \, s^{-1}$ are available on special order from some vacuum manufacturers.

Table 10.2 shows how pumping speed varies considerably for different gases (where nitrogen is taken as 100 per cent). Note that the pumping speeds for helium, argon and neon are small compared with the speed for nitrogen. The chemical inertness of the atoms makes the gettering action relatively ineffective. If the basic diode pump is used to pump significant quantities of air, e.g. in a system with a leak, because air contains approximately 1 per cent argon, a phenomenon known as argon instability will occur. Large periodic rises in pressure to about 10^{-4} mbar will occur due to the release of argon previously pumped, followed by rapid pump-down as the released argon is repumped. The basic diode pump is often modified in some designs by slotting the cathodes to increase the inert gas pumping speed. The slotted cathode presents a surface to a portion of the impinging ions such that glancing incidence and high sputtering rates occur, so that the argon is more likely to be permanently buried. Substantially higher argon speeds can be obtained by using a triode electrode arrangement. In this type of pump (see Figure 10.4) an open titanium structure is positioned between the central anode and the pump body. The new structure is usually in the form of strips and these become the true cathode, the pump body becoming auxiliary electrodes. The pumping mechanisms are exactly the same as in the diode pump but most of the positive ions striking the cathode do so at glancing incidence, producing a substantial increase in the amount of titanium sputtered from the cathode. This leads to enhanced ion burial and noble gas pumping occurring at the auxiliary electrodes. Another triode pump has been produced with cathode plates showing a cell structure with radially arranged fins.

Table 10.2 Variation in pumping speed for different gases in a diode sputter-ion pump relative to nitrogen

Gas	Speed (%)	Gas	Speed (%)
Nitrogen	100	Light hydrocarbons	90–160
Hydrogen	270	Water vapour	100
Oxygen	57	Helium	10
Carbon dioxide	100	Argon	6
		Neon	5

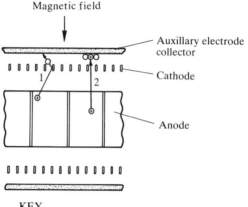

KEY

1   Sputtering of atoms of cathode
    material onto the auxilliary electrode
2   Entrapment of buried argon ions

Figure 10.4   Triode sputter-ion pump

The rate of cathode material usage is proportional to total gas throughput and typically a diode pump has a life of approximately 2000 to 5000 hours at 10^{-5} mbar, 20 000 to 50 000 hours at 10^{-6} mbar and 200 000 to 500 000 hours at 10^{-7} mbar (note that 8800 hours is approximately 1 year of continuous running time).

To start up a sputter-ion pump it is necessary to reduce the pressure to about 10^{-2} mbar, and preferably much lower, by means of a roughing pump. A trapped rotary pump or sorption pump can be used. Sputter-ion pumps can operate in any orientation and do not need water or liquid nitrogen supplies. However, they do have strong magnets which could affect other equipment in the immediate vicinity. They have a long life and can provide ultra-high vacuum, free of organic contamination and vibration. They are used in situations where there is no need to frequently cycle the chamber to atmosphere. They are therefore not generally used in high production applications; instead they are employed mainly for clean surface studies and in those applications where any organic contamination will give unsatisfactory results, such as in research and analytical applications, e.g. in some electron spectrometers and electron microscopes.

10.3 An ion pumped system

Since the sputter-ion pump retains the pumped gases within the pump body it does not need a pump backing it continuously. It will need a roughing pump to reduce the pressure in the system to at least 10^{-2} mbar. A two-stage oil-

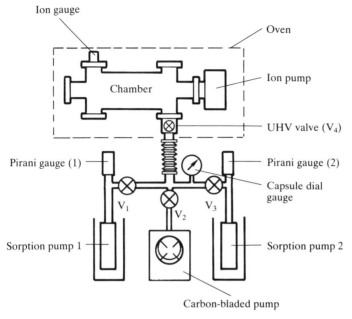

Figure 10.5 Layout of an ion pumped UHV system

Table 10.3 Times and pressures recorded for ion pumps

Condition	Pressure	Time interval
Initial roughing		
1. All valves open. Carbon blade pump used to pump out system (including sorption pumps)	From atmosphere to 120 mbar	30 s
Cool-down		
2. V_1, V_3 closed. Pre-chill of both pumps	Pirani 1 and 2 both read 4×10^{-3} mbar	30 min
Stage 1: sorption pumping		
3. First sorption pump opened. V_2 closed, V_1 open	Down to 1 mbar	10 s
Stage 2: sorption pumping		
4. First pump isolated after 10 seconds by closing V_1; second pump brought in by opening V_3	7×10^{-3} mbar 4×10^{-3} mbar 7×10^{-4} mbar	1 min 5 min 2 h
Stage 3: ultra-high vacuum pumping		
5. Sputter-ion pump started and chamber isolated from sorption pump after 5 minutes. V_4 closed, V_3 closed	5×10^{-9} mbar	After further 16 h
6. Overnight bake of chamber at 200 °C	Less than 10^{-10} mbar	Next day

sealed rotary pump can be used providing it is suitably trapped, using either a liquid nitrogen trap or a sorbent trap. However, since the main attribute of the ion pump is complete freedom from hydrocarbons, ion pumps are often used in conjunction with sorption pumps. In general an isolation valve between the ion pump and system is not needed. A non-lubricated carbon-bladed pump is useful to pre-exhaust the whole system.

The system chamber shown in Figure 10.5 has a surface area of about 0.2 m². Operation of the roughing pumps was carried out in a way similar to that described in Section 10.1. The pump-down times and pressure shown in Table 10.3 were recorded, having first isolated and baked out the sorption pumps to regenerate them. If a titanium sublimation pump had been available in this system and operated at this point, a final ultimate pressure in the low 10^{-11} mbar range would be expected.

10.4 Troubleshooting ion pump systems

Some common fault conditions that occur with ion pump systems are listed in Table 10.4 together with possible causes.

Table 10.4   Possible causes of common fault conditions in ion pump systems

System	Possible causes
1. Slow starting of sorption pumps	(a) Regeneration of sorbent required
	(b) Poor conductance of pipework to sorption pumps
	(c) Air leak
	(d) Improper starting procedure
2. Slow starting of ion pumps	(a) Air leak
	(b) Foreign material in pump, i.e. molecular sieve
	(c) Damaged UHV valve sealing components
	(d) Heavy build up of sputtered titanium*
3. Pressure bursts	(a) Titanium particles flaking from the anode. Pump clean needed (check with supplier)
4. Short circuit	(a) Titanium flake lodged across pump electrodes
	(b) Anode support insulator shorted
5. High leakage currents	(a) Conducting layers on insulators
	(b) 'Whiskers' growing on cathode surface

* Older pumps having undergone long service may exhibit slow starting caused by the build up of substantial sputtered titanium layers. The pump may need to be stripped and cleaned or refurbished. This may entail returning the pump to the manufacturer.

Problems with titanium sublimation pumps

If no increase in pumping speed is achieved when the sublimation rate is increased, an abundance of unsaturated titanium may already be available. Pressure spikes which occur during operation of the pump may indicate that the titanium source is nearly expended.

11

Vacuum pump comparisons

11.1 Rough vacuum to high vacuum

The rotary pump is the most convenient pump for rough pumping duties. It can pump a system down quickly and operate continuously, or pump-down can be repeated again if required. Perhaps one of the possible drawbacks is that the pump uses oil and is therefore a possible source of system contamination through vapour migration. Additionally, the oil may be attacked by certain pumping gas loads. Oil vapour will not migrate from an operational rotary pump against a gas flow from the vacuum chamber providing that suppressive gas flow conditions exist in the backing/foreline or roughing line. These conditions exist down to around 0.1 mbar. If lower pressures are required then oil vapour migration can be suppressed with a foreline trap.

Trapped rotary and sorption pumps

For clean rough pumping, from atmospheric pressure down to around 10^{-3} mbar, either a rotary pump with foreline trap or a sorption pump can be used. A few points comparing their characteristics are summarized in Table 11.1.

Mechanical booster and vapour booster pumps

Table 11.2 compares features of the mechanical booster with the vapour booster pump. In general their peak pumping speeds overlap with the vapour booster, being able to reliably perform at high volumetric speed over the 10^{-1} to 10^{-4} mbar range. The mechanical booster has distinct advantages at higher pressures in the 10 to 10^{-2} mbar range. In general the vapour booster is more able to deal with excessively dirty/dusty applications than the mechanical booster and its exceptional performance pumping light gases (e.g. hydrogen) contrast favourably with the reduced speed for hydrogen exhibited by a mechanical booster.

Table 11.1   Some characteristic comparisons of trapped rotary pumps and sorption pumps

Trapped rotary pump	*Sorption pump*
Hydrocarbon contamination from the rotary pump is suppressed, but not entirely eliminated.	No hydrocarbon-based materials present.
Possible risk of contamination through mal-operation.	No risk of contamination through mal-operation.
Foreline trap must be regenerated to maintain performance.	Molecular sieve requires regeneration to desorb water vapour.
Non-selective. Each component of the initial atmosphere is pumped equally well.	Selective. Neon, helium and hydrogen are not pumped by a single sorption pump. Limited pumping when two pumps are used sequentially.
Continuously operating system.	Cycle operation. After each pump-down the sorbed gases are released by allowing the pump to warm up to room temperature.
A quick pump-down cycle can be repeated immediately.	The number of repeat cycles is limited. Sorbed gases will need to be released and eventual baking of the sorbent will be necessary to remove the retained moisture.
Any size of vacuum chamber can be pumped. Large chambers require longer pumping times or larger pumps.	Multiple sorption pumps operating in parallel are required for large chambers.
No liquid nitrogen.	Liquid nitrogen required to cool molecular sieve.
Transmission of vibration from the rotary pump to the system is limited by means of a clean flexible stainless steel tube.	Vibration free and silent.

Table 11.2   Mechanical booster and vapour booster comparison

	Mechanical booster pump	Vapour booster pump
Best working range (mbar)	$10–10^{-2}$	$10^{-1}–10^{-4}$
Pumping of light gases	Poor	Very good
Pumping of heavy gases	Good	Poor
Ultimate pressure (mbar)	About 1/10 of the backing pump's ultimate pressure	10^{-4}
Effect of dust	Possibly serious	Minor
Power consumption	Power used only when pumping	Power used continuously
Warm-up time (min)	None	25–60
Vapours evolved	None	Back-streams

Vapour booster and diffusion pumps

Table 11.3 compares some characteristics of a vapour booster with a diffusion pump of similar pumping speed; noticeable is the booster's high critical backing pressure.

Table 11.3   Vapour booster and diffusion pump comparison

	Vapour booster pump	Diffusion pump
Working range (mbar)	$10^{-1}–10^{-4}$	$<10^{-3}$
Pumping speed—air (l/s)	4000	4600
—hydrogen (l/s)	6000	5000
Critical backing pressure (mbar)	2–2.6	0.5
Power (kW)	6	3.75
Warm-up time (min)	60	30
Fluid charge (l)	10	$1\frac{1}{4}$
Overall height (m)	1.9	0.75

Comparison of the four pumps

Figure 11.1 compares speed curves for four different pumps, i.e. rotary vane, mechanical booster combination, vapour booster and diffusion pump. We can see that the diffusion pump has a much higher pumping speed compared with the rotary pump that may be used to back it. Note the overlap of the diffusion pump speed curve with the falling speed of the rotary pump between 10^{-2} to 10^{-1} mbar. Note also how the mechanical booster and vapour booster speed curves peak in the region between where the rotary pump and diffusion pump begin to lose speed.

Figure 11.2 shows a vacuum pumping system for an induction melting

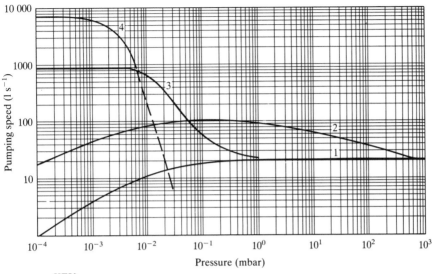

KEY
1 Two-stage rotary pump
2 Two-stage rotary pump and mechanical booster combination
3 Vapour booster
4 Diffusion pump

Figure 11.1   Comparative speed curves for different types of pump

furnace utilizing all of the four types of pump compared in Figure 11.1. There are five identical backing pumping groups, each containing one 2600 m³ h⁻¹ mechanical booster pump backed by two single-stage 275 m³ h⁻¹ rotary pumps. Additionally can be seen the conical water-cooled bodies of two vapour boosters and partly hidden two 20 in diffusion pumps. Such combinations provide fast roughing times and high pumping speed over the whole pressure range from atmospheric pressure to below 10^{-4} mbar.

11.2 High vacuum to ultra-high vacuum

Diffusion pumped systems

Useful features
- Simple (reliable)
- High speed
- High throughput
- Low cost
- Withstands contamination
- Range of speeds available 10–90 000 l s⁻¹
- High speed for water vapour when fitted with a liquid nitrogen trap

Poor features
- Back-streams, needs baffle (note that integrated systems do not)
- Needs rotary pump constantly
- Vertical orientation only

Figure 11.2 Vacuum pumping system for an induction melting furnace

Pumping speed is comparatively non-selective although there is some relation with molecular weight, e.g. H_2 is pumped some 20 per cent faster than nitrogen. The pump has no 'memory effects' from previously pumped gases.

System operation is a straightforward multistep procedure but with the opportunity for serious mal-operation on manual systems, which could result in system/chamber contamination with diffusion or rotary pump oil vapours. A 'clean' residual gas atmosphere can be obtained using low vapour pressure oils such as polyphenyl ether and a suitable baffle. A liquid nitrogen trap and metal seals are needed to reach ultra-high vacuum.

A typical residual gas composition of a system pumped to UHV with a diffusion pump is shown below:

H_2	CH_4 plus H_2O	CO plus N_2	CO_2
10%	15%	65%	10%

The gas composition to a great extent depends on the design and construction of the process vessel (in this and other examples given here the systems had metal seals). Section 4.5 lists possible sources of these residual gases.

The pump can take 15 minutes or more to achieve operating temperature—depending on size. Once operating, it can pump rapidly and is often used for rapid-cycle processing. Such applications will require a high vacuum isolation valve.

Turbomolecular pumped systems

Useful features
- Fully pumping speed obtained quickly
- Low operating costs
- Clean
- Choice of orientation (some models)
- Can be valveless
- Range of speeds available 50–9000 l s^{-1}

Poor features
- High capital cost
- Separate controller needed
- Low compression ratio for light gases
- Susceptible to damage
- Difficult to maintain
- Needs rotary pump constantly

The turbomolecular pump is a very versatile pump and simple to use; generally only an on/off push button on the controller is used to initiate and end the pump-down. It is suitable for rapid cycling plant and systems can be valveless. Hydrocarbon-free residual atmospheres can be obtained provided that:

1. Air is admitted correctly so that lubricant vapours are not 'blown' towards the vacuum chamber.

2. The pump is not stopped and left evacuated otherwise oil/grease vapour from turbomolecular pump bearings and backing pump will back-migrate towards the chamber.

Essentially it pumps most gases with constant pumping speed and without memory effect. It does not require liquid nitrogen. The pump must be vented when stopped. The pump produces high-frequency vibration but this is generally negligible when compared with that originating from the rotary pump. The lowest pressure obtainable is limited by the partial pressure of hydrogen. This is due to the poor compression ratio of turbomolecular pumps for light gases. Efficiency is greatest for heavy gases and poor for H_2 and He. These light gases can back-diffuse through the pump. Turbopumps are sometimes backed by diffusion pumps to overcome this. Typical residual gas composition of a system pumped with a turbomolecular pump to UHV is shown below:

H_2	CH_4 plus H_2O	CO plus N_2	CO_2
50%	10%	35%	5%

Recent designs have powerful motors for rapid acceleration to optimum speed. This allows pumps to be used on rapid-cycle time plant. Pumps are also available that can deal with corrosive or radioactive gases.

Certain models of turbomolecular pump may be mounted in any orientation. The disadvantages of turbomolecular pumps, however, include high initial cost, compared with diffusion pumps, and their susceptibility to damage.

Cryopumped systems

Useful features
- High speed for water vapour
- High throughput
- Very clean
- Low operating costs
- Rotary pumps only required to rough out
- Choice of orientation
- Range of speeds available 400–18 000 l s^{-1}
- Fail-safe when mal-operated— no contamination

Poor features
- Requires regeneration
- Requires periodic maintenance
- High capital cost
- Low capacity for hydrogen and helium

Cryopumps are 'clean'. They operate by freezing out the gases being pumped; helium, hydrogen and neon are pumped by sorption. The pumped gases are trapped within the pump and can be released if the pump warms up due to power failure. Some stored gases may have highly undesirable properties. The pump usually has a built-in pressure relief safety valve. Pump efficiency reduces as condensed gases build up. It is good for all gases except He, H_2 and Ne. Cool-down time can be extensive (typically 1 to 3 hours). Time is also required for regeneration. Services needed are mains power and water for the compressor. Pumps can be used in any orientation.

Sputter-ion pumped systems

Useful features
- Ultraclean
- Low operating costs
- Needs backing pump to rough only
- Choice of orientation
- Pump can be baked
- Power failure not a disaster
- Requires only electrical supply

Poor features
- Low speed and throughput
- High capital cost
- Speed varies with gas species
- Magnetic field present

Sputter-ion pumps are not generally used for quick-cycle, high-speed production systems but have an important role in the UHV field. It is a completely hydrocarbon-free and vibrationless pump. A high vacuum isolation valve is not normally required. It exhibits pronounced selective pumping of gas species, together with a memory effect resulting from liberation of previously pumped gases by ion bombardment. Typical residual gas composition of a system pumped to UHV is shown below:

H_2	CH_4 plus H_2O	CO plus N_2	Ar	CO_2
15%	10%	60%	5%	10%

The pump is not suitable for inert or hydrocarbon vapour loads (hydrocarbons are dissociated in the discharge and carbon is deposited on the cathodes). The discharge intensity (and therefore pumping speed) falls with pressure.

The absence of vibration makes it very well suited for use on electron microscopes and surface analysis equipment. Its cleanliness finds it widely employed in mass spectrometry and electron tube pumping. There are some applications where the stray magnetic field from the sputter-ion pump may cause problems in the process vessel. Stray fields are reduced by increasing the distance between the ion pump and process vessel, by incorporating an elbow so that the axial field is not directed into the process vessel and by fitting magnetic shielding (overlapping mild steel sheets bolted tightly together) around the pump.

General comments

All high/ultra-high vacuum pumps operate in the molecular flow region. The pumps must be rough-pumped before they can be started. High vacuum pumping systems for general-purpose or fast cycling duties would most likely employ diffusion or turbomolecular pumps. These would be preferred because they both have good high throughput pumping characteristics, and the gas load is exhausted to the atmosphere and the pump therefore has no memory effects. For 'dirty' conditions the diffusion pump would be the first choice.

Diffusion and turbomolecular pumps exhibit a constant throughput characteristic at pressures above 10^{-3} mbar and a constant pumping speed below 10^{-3} mbar. Eventually, pumping speed will fall as a consequence of the finite compression ratio for the gas being pumped. This effect is noticeable when pumping a hydrogen load, particularly with turbomolecular pumps. Pumping speed is re-established to a lower fine-side pressure if lower backing pressures are provided, e.g. with series-connected diffusion or turbomolecular pumps.

The pumping speed of a sputter-ion pump depends on discharge characteristics and titanium sputtering rates. The rated speed for nitrogen is low at high pressures, rises to a maximum between 5×10^{-6} mbar and 1×10^{-6} mbar and falls to 50 per cent of the maximum speed between 1×10^{-9} and 5×10^{-10} mbar. The pumping speed of a cryopump at a particular time depends on the temperature of the cryopumping surfaces and the degree of saturation of the charcoal surfaces (particularly with hydrogen). Pumping speeds for cryopumps are therefore not plotted as a function of pressure.

Diffusion, turbomolecular and cryopumps are essentially non-selective: all gases are pumped equally well. Sputter-ion pumps exhibit selective pumping and pumping speed depends on the chemical nature and composition of the gas load.

Turbomolecular pumps might be chosen for ease of operation, fast starting, mounting flexibility or because of their ability to recover from accidental 'air-dumping'. On the other hand, diffusion pumps may be chosen for ease of maintenance, low vibration/noise or the ability to withstand debris falling into the pump. Turbomolecular or cryopumps might be chosen where the process may be sensitive to possible contamination from diffusion pump fluids. In practice, it has been reported that surface analysis systems have been built using diffusion, turbomolecular, ion and cryopumps and, where similar surfaces have been investigated in systems employing different types of pump, sufficiently good agreement has been obtained to indicate that in properly designed systems, the state of the surface under study is not significantly affected by any contribution of the pumping system to the background gas.

Since water is a primary residual gas and would be a process contaminant in semiconductor applications, this makes cryopumps attractive on fast cycling equipment that must reach low base pressures quickly. Vibration from cryopumps and turbomolecular pumps may be unacceptable for some applications but can be effectively suppressed by commercially available dampers. Capital costs and running costs are also important considerations. Figure 11.3 gives a comparison of capital costs for the high vacuum pumps considered. It should be emphasized that comparisons of this type will inevitably be different from country to country and will change with time as

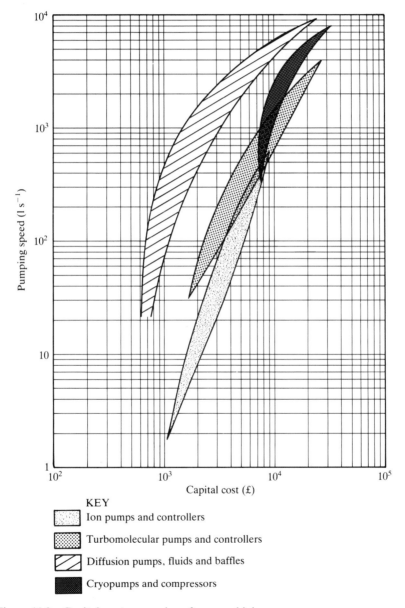

KEY

[ ] Ion pumps and controllers

[ ] Turbomolecular pumps and controllers

[ ] Diffusion pumps, fluids and baffles

[ ] Cryopumps and compressors

Figure 11.3   Capital cost comparison for some high vacuum pumps

prices vary. Note that capital costs of complete systems for true comparison may include high vacuum valves for diffusion and cryopumps and possibly liquid nitrogen traps for diffusion pump systems, etc. However, it can be seen that the diffusion pump, with fluid (polyphenyl ether) and baffle, is always cheapest and ion pumps with their associated controller (up to $700 \, l \, s^{-1}$) are the most expensive. The diffusion pump/fluid/baffle combination costs roughly one-fifth that of an unvalved turbomolecular pump with frequency converter.

The greatest part of the cost of a cryopump lies in the compressor; consequently for reasons of economy cryopumps are not available in sizes less than around $400 \, l \, s^{-1}$. However, it should be noted that some compressors are capable of running two cold heads.

Running and service costs are also important considerations when choosing a pump. Even taking into consideration the fact that they do not always require a rotary pump, cryopumps are usually more expensive to run than diffusion and turbomolecular pumps.

Turbomolecular pumps use about 25 per cent of the power of a comparatively sized integrated vapour pumping group and about 30 per cent of the water consumption.

More detailed comparisons should take into account shut-down times for regeneration of cryopumps, liquid nitrogen costs for liquid nitrogen traps if included in diffusion pump systems, etc. If traps are included then the diffusion pump is considerably more expensive to operate than any of the other pumps.

Ion pumps are the least expensive high vacuum pumps to operate. Below 10^{-7} mbar the power consumption is negligible. For example, a $150 \, l \, s^{-1}$ ion pump power supply delivers only $100 \, mW$ at a pressure of 10^{-8} mbar compared with a diffusion pump of the same speed having a $0.4 \, kW$ heater. However, power consumption of the ion pump system's heaters during bake-out periods should be taken into account.

Cryopump maintenance is normally limited to the compressor and cold head seals and is comparatively easy compared to bearing changes on the turbomolecular pump. Diffusion pumps will run almost indefinitely without maintenance. The only components liable to replacement are the heater and oil, and this can be easily accomplished. Ion pumps do not require routine maintenance apart from periodic checking and cleaning of the high-voltage feedthrough and the high-voltage cable and its connector.

In conclusion, there is no single ideal pump which will evacuate from atmosphere to 10^{-10} mbar. However, commercially available vacuum pumps used in a suitable combination allow high and ultra-high vacuum pressures to be achieved readily and consistently. Some of these combinations approach the ideal in some aspects of their performance. The choice of pumps to a great extent is determined first by the characteristics of the vacuum process and second by the characteristics of the pumps.

12

Vacuum system connections, components and assembly

12.1 Demountable joints

Demountable joints for vacuum in the range from atmospheric pressure down to 10^{-7} mbar normally utilize 'O' rings for the sealing method. The 'O' rings are made of an elastomer material which is trapped and compressed between two surfaces to form the seal. Compression can be achieved by bolted, screwed or clamped flanges.

Demountable flange joints

Figure 12.1a shows the 'O' ring located in a groove machined in one of the flanges. Here the groove section is trapezoidal which has the advantage of better retention than a plain rectangular groove. The cross-sectional area of the groove should be greater than that of the 'O' ring so that when the second plain flange is bolted on, the 'O' ring is distorted until it almost fills the groove. With all types of joints it is essential that the surfaces of the groove and the plane flange should be free of scratches or 'foreign objects' such as hair, swarf, etc., which could act as leakage paths.

Figure 12.1b shows a trapped 'O' ring carrier suitable for thin flanges where grooves cannot be machined. This carrier usually consists of an inner and outer circular metal ring with the 'O' ring trapped in between. Under compression the two flanges may be in contact with the metal carriers, and the inner ring can then act as a potential barrier to outgassing vapours from the 'O' ring, making their journey towards the vacuum system very tortuous. Under extreme conditions where the barrier can be considered as virtually complete (this may be enhanced if pumping fluid reaches the area), gas will accumulate inside the trapped volume so formed. In some cases the gas pressure will increase until its pressure is great enough to overcome the obstruction, when the gas will 'burst' through into the vacuum. This behaviour can be observed by the associated system vacuum gauges as a very sharp characteristic pressure pulse appearing at regular intervals (see Figure 12.2).

A newer gasket construction is shown in Figure 12.1c. Here the sealing ring is directly exposed to the vacuum system, being held by a specially designed

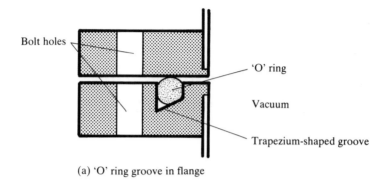

(a) 'O' ring groove in flange

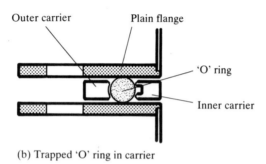

(b) Trapped 'O' ring in carrier

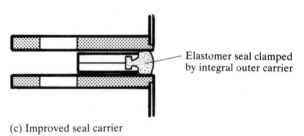

(c) Improved seal carrier

Figure 12.1 Various demountable flange joints

plastic carrier on the atmospheric pressure side only. The problem of gas bursts is thus removed.

Various flange assembly arrangements exist and some examples are shown in Figure 12.3. These consist of:

1. *Bolted flanges,* where the flange is an integral part of the component body.
2. *Bolted rotatable flanges and collars.* Here only the sealing face (collar) is an integral part of the component body. The rotatable part of the flange is

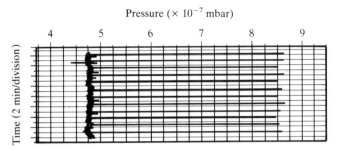

Figure 12.2   Typical pressure record when pressure bursts are present

held in place by a circular clip. This interchangeable construction has advantages for products such as diffusion pumps and other components which may have to mate with flanges of different sizes and fixing arrangements.

3. *Clamped flanges using collars only.* This arrangement can be used where collars of the same size are to be used.

Figure 12.4 shows an example of how mixed flange arrangements can be assembled. Flanges like those illustrated have a bolt hole circle and other dimensions corresponding to International Standards (i.e. ISO).

Grooved flanges

The recommended type of grooved seal is of the limited compression type, to ensure constant deflection of the 'O' ring and to limit the compression of the

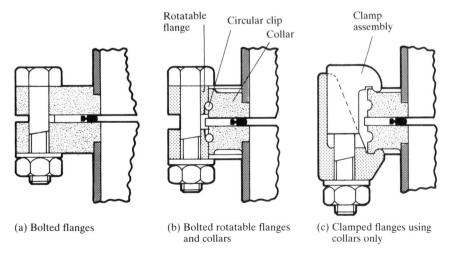

(a) Bolted flanges

(b) Bolted rotatable flanges and collars

(c) Clamped flanges using collars only

Figure 12.3   Various flange assembly arrangements

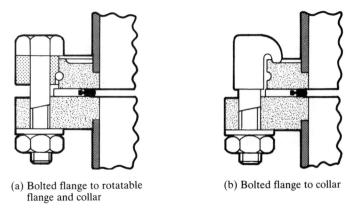

(a) Bolted flange to rotatable flange and collar

(b) Bolted flange to collar

Figure 12.4 Assembled mixed flange arrangements

'O' ring to that required for a vacuum-tight seal. When this compression ratio is reached, the flanges meet in a metal-to-metal contact, enclosing the 'O' ring in the groove.

Such grooves should meet the following requirements:

1. The cross-section of the groove should be of the order of 10 per cent greater than the cross-section of the 'O' ring, to allow for a metal-to-metal flange contact without overcompressing the 'O' ring.
2. To provide positive retention of the gasket, when the seal is not 'made'.
3. To provide the minimum volume of trapped gas.
4. To allow reuse of the 'O' ring after the seal is 'broken'. In this respect, overcompression or grooves with sharp corners and/or acute angles are of

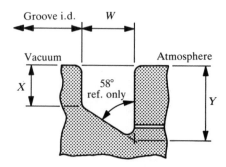

Radii of corners:
0.005–0.010 in (0.13–0.25 mm)

Radius of acute angle in groove base:
0.020–0.030 in (0.51–0.76 mm)

Figure 12.5 Trapezium grooves

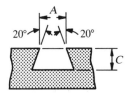 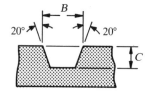

(a) Closed dovetail
Demountable vacuum joint
with dovetail groove for
positive retention of the
'O' ring

(b) Open dovetail
A similar joint without
positive 'O' ring retention,
but easier to machine.

Tolerance
A, B and $C \pm 0.002$ in (0.05 mm)

Radii of all corners:
0.005–0.010 in (0.13–0.25 mm)

Figure 12.6 Dovetail grooves

bad design as this leads to permanent deformation or cutting of the 'O' ring.

5. To permit easy flange machining and wide dimension tolerances.

Two main configurations of grooves are in common use in vacuum seals, trapezium and dovetail grooves as illustrated in Figures 12.5 and 12.6 respectively. Dimensional details of each type are given in Tables 12.1 and 12.2.

In order to avoid the difficult machining needed for the dovetail groove, as well as to eliminate the trapped volumes formed in these grooves, the trapezium groove with parallel sides is preferred. In this groove the 'O' ring completely fills the vacuum side, a very small volume of gas being trapped on the atmospheric side. A small hole in the flange leading from the groove to atmosphere is shown. This allows the assembled joint to be leak-tested by the injection of tracer gas.

Note that the surface finish for sealing flange faces should be of the order of 1.5 μm C.L.A. (Centre Line Average) (0.000 06 in) for 'O' rings.

In situations where non-standard 'O' rings are required for demountable flange joints, these can be made from extruded elastomer cord. The joint between the ends can be made with or without glue.

Without glue the length of the cord must be about 5 per cent greater than the mean 'O' ring circumference, to allow adequate compression for sealing. The ends must be cut square. When using glue such as 'Superglue', the joint will be more rigid than adjacent areas of elastomer. A glued butt joint is therefore not normally reliable since it is likely to crack. A scarf joint must be used (see Figure 12.7). The orientation of the joint must be as shown in illustration (a) not as in (c), since this again will be rigid and liable to crack.

Table 12.1 Dimensional details of trapezium grooves

Dimensions (*mm*)

'O' ring section 'd'		W	X	Y
1.78	max.	1.68	1.19	2.21
	min.	1.63	1.14	2.16
2.62	max.	2.51	1.65	3.18
	min.	2.44	1.57	3.10
3.53	max.	3.40	2.18	4.27
	min.	3.30	2.08	4.17
5.33	max.	5.18	3.23	6.38
	min.	5.03	3.07	6.22
6.99	max.	6.81	4.19	8.31
	min.	6.60	3.99	8.10
7.92	max.	7.70	4.83	9.50
	min.	7.47	4.60	9.27
12.70	max.	12.42	7.57	15.09
	min.	12.04	7.19	14.71

Table 12.2 Dimensional details of dovetail grooves

Dimensions (*mm*)

'O' ring section 'd'	A	B	C
1.78	1.50	2.41	1.24
2.62	2.34	3.60	1.88
3.53	3.07	4.93	2.54
5.33	4.70	7.44	3.78
6.99	6.17	9.93	5.16
7.92	6.68	10.92	5.84
12.70	11.30	18.16	9.42

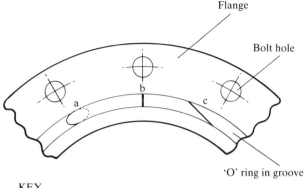

KEY

a 'Scarf' joint acceptable
   when using glue
b 'Butt' joint acceptable
   when no glue is used
c Non-acceptable scarf joint

Figure 12.7   Non-standard 'O' ring made from elastomer cord in an 'O' ring groove showing different types of joints

'O' rings should never be removed with a metal tool, such as a screwdriver, as this can inevitably damage the groove and flange surfaces, possibly leading to a potential, if not an actual, leak. Instead it is recommended that a wood or plastic tool be used. The 'O' ring can be removed by inserting the tool (such as a toothpick or thin piece of plastic card) between the outside of the 'O' ring and groove, and sliding this tool around the 'O' ring. This causes the 'O' ring to 'pop up'. It may be necessary to hold the 'O' ring down on the side opposite to the tool to prevent the 'O' ring from turning in the groove. This technique gives better extraction than trying to pry the 'O' ring from the groove.

Demountable pipeline couplings

Several forms of demountable pipeline couplings have been produced in the past, all of them using an 'O' ring type of seal. The one most widely used at present and available from many vacuum component manufacturers have small flange terminations which conform to ISO, Pneurop and British Standards in terms of size availability and joint geometry. Figure 12.8 shows a selection of components which are available in a range of sizes and include elbows, tees, crosses, flexible connections, etc. Figure 12.9 illustrates the profile and size ranges of flange terminations. The number designating the coupling size refers approximately to the bore diameter of the associated tube, i.e. a KF40 or NW40 coupling is used with a 40 mm bore tube. Note that NW is an abbreviation for the German word *nennweite* (nominal width) and KF for *klein flange* (small flange).

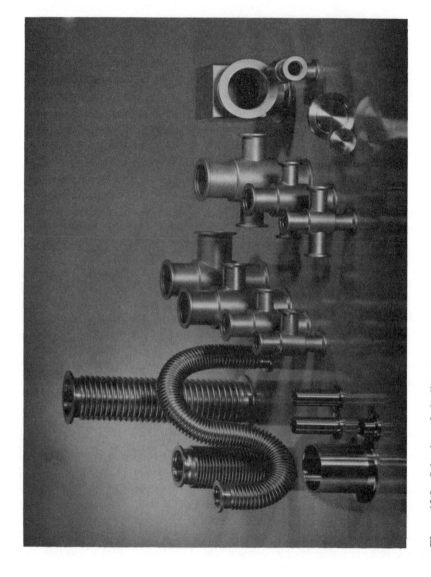

Figure 12.8 Selection of pipeline components

NW profiles

	A	B	C	D
NW10	30	12.2	2	14
NW16	30	17.2	2	20
NW20	40	22.2	2	25
NW25	40	26.2	2	28
NW32	55	34.2	2	38
NW40	55	41.2	2	44.5
NW50	75	52.2	3.2	57

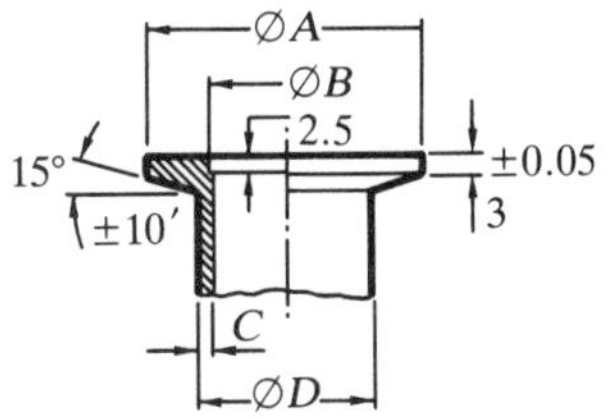

Figure 12.9   Profile and size of small NW flange connections (dimensions in mm)

In order to be able to connect two components, a centring ring (made of metal or plastic) holding an 'O' ring on its outside periphery and a clamping ring are necessary. Figure 12.10 shows how a joint can be made, without tools. Where the joint is likely to be internally pressurized, this arrangement for holding the 'O' ring may not be satisfactory and the seal may leak. An alternative type of carrier for the 'O' ring similar to that shown in Figure 12.1c is available and suitable for pressures up to 17 bar. Hinged clamps are available for quick-release applications where a joint is to be made and taken apart frequently. Figure 12.11 shows how a typical hinged clamp closes.

Precautions in handling and cleaning of 'O' rings

'O' rings are capable of absorbing large quantities of the solvents used in cleaning vacuum components and must *not* therefore be cleaned in this way. Subsequent evaporation of the solvent within the vacuum system would be a great problem, producing large quantities of vapour to be pumped away.

If flange assemblies and 'O' ring carriers need to be degreased the 'O' rings should be removed and treated separately by carefully wiping clean with a lint-free material. The metal parts can be degreased in the normal manner. If 'O' rings are found to be non-flexible, have a permanent set or have surface damage they should be replaced.

Static seals, where possible, are assembled without the aid of vacuum oils or greases; however, *very* light lubrication is acceptable in certain circumstances (except in ultra-clean applications). The function of oil or grease on an 'O' ring is to lubricate; in this context a light lubrication on assembly to help slip the 'O' ring into the groove can be useful. Oils and greases should never be used as a means of filling sealing surface imperfections. When used the quantity of oil/grease should be sufficient only to give the elastomer a sheen. Excessively large quantities can lead to the accumulation of dirt and grit, etc., which can potentially lead to leakage problems.

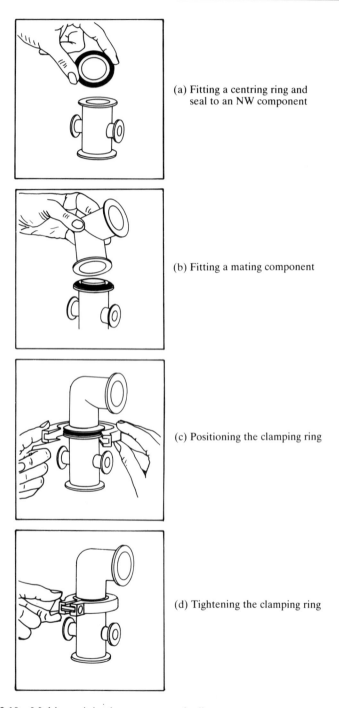

(a) Fitting a centring ring and seal to an NW component

(b) Fitting a mating component

(c) Positioning the clamping ring

(d) Tightening the clamping ring

Figure 12.10 Making a joint between two pipeline components

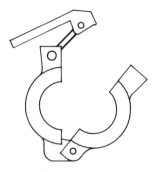

(a) Clamp open

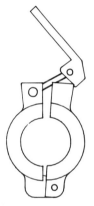

(b) Clamp part closed

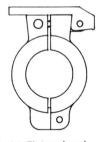

(c) Clamp closed

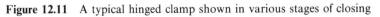

Figure 12.11   A typical hinged clamp shown in various stages of closing

Lubricants are required where 'O' rings rub against mating surfaces, e.g. in quarter-swing valves, shaft seals, etc. In such dynamic situations hydrocarbon, Fomblin or silicone grease is recommended, the latter two being preferred as there is less tendency for the 'O' ring to stick after long static periods. Note that although silicone grease is recommended for dynamic 'O' ring use it does not lubricate metal bearings and must not be used in this application. Some details regarding the characteristics of the greases mentioned are listed in Table 12.3.

Table 12.3 Some characteristics of vacuum greases

	Apiezon type (hydrocarbon)	Silicone type	Fomblin
Good lubricant	Yes	No	Yes
Vapour pressure (mbar)	Low	Approx. 10^{-6}	Low
Temperature suitability	125 °C max.	-40 to 200 °C	-20 to 200 °C
Cost	Moderate	Low	High
Chemical stability	Moderate	Good	Very good

Demountable UHV joints

The main requirement for UHV joints is that they have a low outgassing rate and leak-tightness to less than 10^{-10} mbar l s^{-1}. In general, therefore, metal gasket materials are preferred, either of wire (see Section 13.6) or a flat annular disc such as the copper ConFlat® flange. This type is the most universally used.

The ConFlat® flange (see Figure 12.12) has a knife-edge sealing-type geometry. This design has built-in long-term reliability because it captures the gasket, preventing it from flowing away from the seal area even with baking temperatures of 450 °C and thus maintaining sealing forces. As the knife edges begin to bite into the copper gasket, the gasket material flows outward. The outer rim of the flange restricts the flow. It is important that the knife edges are protected from mechanical damage when not in use.

The following points should be noted:

1. Clamping must be uniform with many bolts, closely spaced.
2. Initial compression of the gasket must be done in a controlled sequence. Use thread lubricant on bolts to ensure easy dismantling.
3. During baking, heating must not relieve sealing load, either by movement of gasket material or stretching of bolts. The rate of temperature rise should be less than 150 °C per hour.

It is not advisable to reuse copper gaskets.

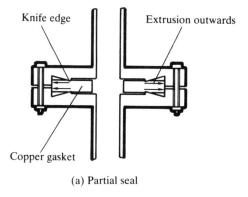

Knife edge   Extrusion outwards

Copper gasket

(a) Partial seal

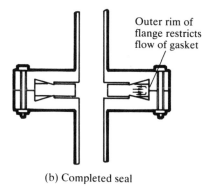

Outer rim of
flange restricts
flow of gasket

(b) Completed seal

Figure 12.12   ConFlat® flange sealing arrangement. (ConFlat® is a registered trademark of Varian Associates)

After manufacture, the gaskets are chemically etched to produce a scratch-free sealing surface. Stains on copper gaskets from cleaning processes are not detrimental to seal performance or total system performance. Metal polish must not be used to remove gasket tarnish. They should be dust free and should not be handled without gloves.

Gaskets are manufactured from oxygen-free high-conductivity (OFHC) copper. Special purpose gaskets are available, such as annealed copper gaskets for circumstances where even small distortions of flanges must be avoided. Silver-plated gaskets are recommended for use on vacuum systems that will be subjected to long periods of baking; otherwise heavy external gasket oxidation can occur. Such oxidation in itself presents no problem; however, there is a risk that on dismantling a joint, this loosely adhering oxide scale could fall into the vacuum system or on to the flange sealing area.

Sometimes gaskets are difficult to remove, particularly after extended high-temperature bakeout. Flanges can be separated by carefully levering them apart with a probe (scribe) inserted along the leak-test groove normally provided in the flange. A method of removing the gasket is to compress the part of the gasket that protrudes above the flange in the jaws of a vice. The gasket bows and comes clear.

12.2 Vacuum valves

Valves are normally a necessary part of every vacuum system. They are required to isolate either part or all of the pumping group from the system which is being evacuated or are necessary to control the pumping sequence. Additionally other valves are required to admit air (or gas) into a system.

The main parameters which must be considered when choosing valves for vacuum systems are the working pressure, the leak-tightness of the valve, its conductance properties and the outgassing rates of the materials of construction of the valve.

Valves for rough to medium vacuum

Probably the most popular type of valve in the rough to medium vacuum range is the diaphragm-sealed isolation valve. The principle of operation of this valve is that a flexible elastomer diaphragm is deformed and sealed onto a polished seat (see Figure 12.13). The mechanism is isolated from the system by the diaphragm and is thus unaffected by dirt and other contaminants from

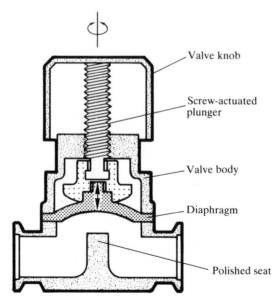

Valve knob

Screw-actuated plunger

Valve body

Diaphragm

Polished seat

Figure 12.13 Diaphragm valve

within the vacuum system. The valve body is normally manufactured from aluminium. Diaphragms are usually available in Nitrile and fluoroelastomer. Fluoroelastomers, which have a higher chemical resistance than Nitrile, are normally used in applications where the valve may be exposed to oils, lubricants, mineral acids, chemicals or solvents. This type of valve is therefore an extremely rugged and system-tolerant valve. Valves of this type normally terminate in flanges suitable for connection to other ISO standard sealing connections.

Valves for medium to high vacuum

Both outgassing and the conductance properties of valves become important in the medium to high vacuum region. These factors affect the ultimate pressure and/or the pump-down time of the system. Four suitable valves will be considered:

1. Bellows valves
2. Quarter-swing valves
3. Gate valves
4. Baffle valves

Bellows valves
Figure 12.14 shows a right-angle plate valve with the transmission enclosed by stainless steel bellows. The seal is made by an elastomer 'O' ring gasket. As

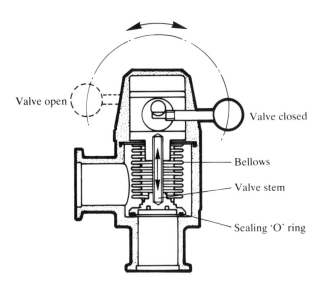

Figure 12.14   Right-angled bellows valve

with most high-vacuum valves the 'O' ring groove is vented to prevent gas bursts from trapped gas. Such valves can be manually, pneumatically or electrically operated. Similar valves which combine the function of separate backing/roughing valves in one integral three-port unit are also available. They are normally used with vapour pumping groups (Section 7.9).

Quarter-swing valves
Quarter-swing valves (see Figure 12.15) have a flat circular plate which swings through a right angle to provide high conductance. The seal is made by an 'O' ring situated in a groove around the perimeter of the plate. Quarter-swing valves are capable of resisting atmospheric pressure differentials in either direction and can either be used as pipeline valves or incorporated into a high vacuum pump system to form a high vacuum isolation valve. When selecting these valves for combination with pumps, baffles, traps, etc., the dimension of the valve plate becomes important, since it can protrude beyond the valve body on opening and could be prevented from opening fully by contact with the internal ports of the component it is attached to.

Gate valves
In the gate valve the valve plate moves in a plane perpendicular to the common axis of the inlet and outlet ports. In the open position the plate is completely clear of the valve bore. Gate valves thus have a slightly higher conductance than quarter-swing valves. Gate valves are necessary if energetic particles must be beamed through the valve. In the valve shown in Figure 12.16, the valve is closed by movement of the carriage. The carriage runs inside the valve body, constrained by four guide wheels until two rollers hit the end of body and prevent further forward movement of the gate. The carriage, however, continues to move forward and forces the plate to seal against the flange by means of two short links.

On opening the valve, a single spring ensures that the link mechanism is reversed and the plate is lowered before it has moved into the valve body, hence minimizing 'O' ring drag. Several different sizes of gate valve are shown in Figure 12.17.

Baffle valves
Figure 12.18 shows a manually operated baffle valve. Here the valve plate acts as a baffle when the valve is open. The plate moves back, on opening, so that its lower face remains parallel with the plane of the orifice. If used above a diffusion pump it therefore acts as a baffle to oil vapour back-streaming. For cooling, it relies on thermal conduction through supports to the valve body which can be water cooled. In another version of baffle valve the ports are at right angles and the valve plate tilts through 45°.

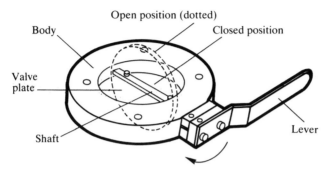

Figure 12.15   Quarter-swing valve

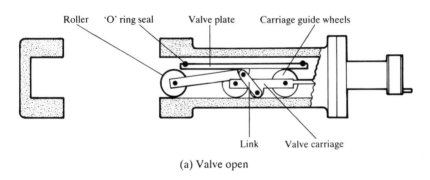

(a) Valve open

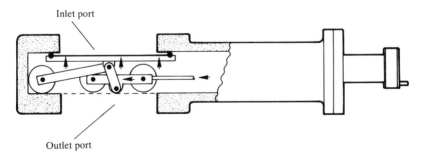

(b) Valve closed (carriage moves sideways and the valve plate up)

Figure 12.16   Gate valve

Figure 12.17   Some examples of stainless steel gate valves with bore sizes from 40 up to 320 mm

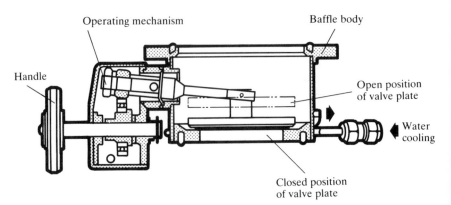

Figure 12.18   Baffle valve

UHV valves

Valves for UHV applications must have very high conductance, low outgassing rates and maximum cleanliness. Most manufacturers use stainless steel as the main body material. In addition, vacuum brazing techniques are employed to ensure that crevices do not appear in any joints. UHV vacuum valves are often baked in order to desorb any gases from the internal surfaces, thereby reducing outgassing. Fluoroelastomers can only be heated to about 200 °C. In cases where it is necessary to bake the valve to about 450 °C, metal-to-metal seals are used. The usual arrangement, similar in construction to the bellows valve, utilizes a stainless steel knife edge and a soft copper or gold-plated copper pad against which the knife edge is forced to form the seal. Precision guiding of the mechanism is required, to ensure that the sealing components always come together in exactly the same position to give many-cycle reliability. As this type of valve can easily be damaged by grit or excessive torque, the valves are designed so that the seals can be easily replaced.

Air (or gas) admittance valves

A simple design of air admittance valve is shown in Figure 12.19. The valve incorporates a control knob attached to a screw-actuated plunger which is sealed by a non-rotating 'O' ring onto a seat in the body. Similar types of air or gas admittance valve can be electrically operated having a solenoid-activated plunger operating against a spring in the housing assembly. The valve would normally be open to atmosphere when de-energized.

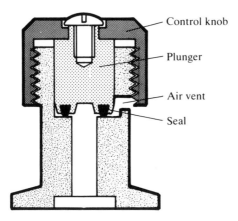

Control knob

Plunger

Air vent

Seal

Figure 12.19 Coarse air admittance valve

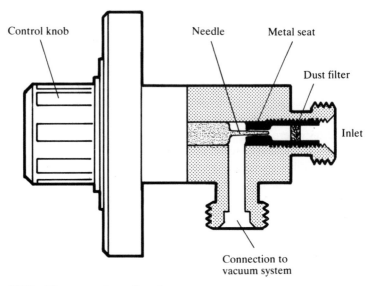

Figure 12.20   Fine-control needle valve

Fine control needle valves

In some vacuum applications, it is necessary to provide a fine control of gas bleed into a vacuum chamber, perhaps to regulate the pressure in the system. Such control is usually provided by needle valves. Figure 12.20 illustrates an extra-fine control valve in which the needle is moved into a metal seat. The movement is transmitted through an 'O' ring rotary seal. It is important that no dirt enters the valve, and the use of porous sintered in-line filters on the inlet side of the valve is recommended.

12.3 Vacuum leadthroughs

A wide variety of standard leadthroughs are available. The range includes rotary and linear motion leadthroughs, multipin and multitube leadthroughs for thermocouples, high current, high voltage, water and liquid nitrogen, etc. Some examples are given below.

Electrical leadthroughs

A selection of electrical leadthroughs are shown in Figure 12.21. They provide a vacuum-tight means of conveying electricity supplies to the inside of a vacuum chamber. A shield is provided to prevent the deposition of metallic coatings on the ceramic insulator during material evaporation in coating applications. The leadthroughs are primarily designed to be 'O' ring sealed into specially shaped baseplate holes.

Figure 12.21   Selection of electrical leadthroughs

Motion leadthroughs

Rotary movement can be transmitted into a vacuum chamber from an external atmospheric pressure source by rotation of a highly polished shaft through the wall of the chamber. A leak-tight seal is made around the shaft using an elastomer material, i.e. either an 'O' ring or some form of lip seal. Depending on the application, with shaft sizes larger than 3 mm it is usual to provide a means of lubrication. An example which is grease lubricated is shown schematically in Figure 12.22; here two 'O' rings are used (the space between them being packed with a low vapour pressure grease). The double 'O' ring seal provides better isolation between the vacuum chamber and atmosphere. Longitudinal movement is also possible. In cases where low outgassing rates are required these seals would be unsuitable and a 'wobble'

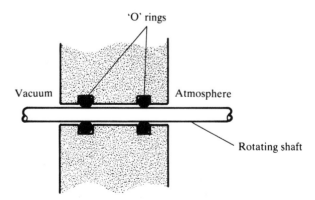

Figure 12.22   Rotary shaft seal

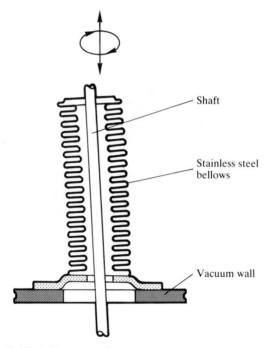

Figure 12.23 'Wobble' drive via bellows

drive is frequently used. Here, rotary and linear motion is transmitted to the vacuum by circular bending motion of a stainless steel bellows with a shaft projecting through the closed end (see Figure 12.23).

Apart from elastomer and metal-sealed motion leadthroughs, an alternative more recent development has been with the use of ferrofluids to make leak-tight seals. A ferrofluidic seal is a suspension of submicroscopic magnetic particles in a low vapour pressure fluid. In these fluids the magnetic particles do not settle to the bottom of the fluid or float to the top, but instead are permanently suspended uniformly throughout the fluid. Figure 12.24 shows a basic ferrofluidic seal; here the fluid fills the gap between the special-shaped pole pieces and the shaft to achieve sealing with no rubbing surfaces or mechanical wear.

12.4 Fabrication of vacuum components

Given below is some practical information for those workers who are prepared to assemble their own vacuum system from the main components. This may involve fabrication of the vacuum chamber together with fixtures and fittings, gauge connections, pipelines, etc.

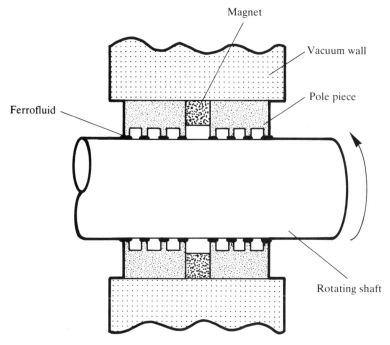

Figure 12.24   Ferrofluidic shaft seal

Materials

Factors to consider when selecting materials for vacuum use include:

1. Vapour pressure at working temperature in relation to process pressure
2. Outgassing, moisture absorption and permeability characteristics
3. Sensitivity of materials to operation hazards, i.e. degradation due to corrosion, radiation, temperature, etc.
4. Mechanical strength and ease of machining or fabricating

The following types of material and surface are to be avoided if possible:

1. Porous materials, because of liability to leakage and retention of absorbed gas.
2. Rough-textured materials, because this often indicates surface porosity. The gas-absorbing ability of a rough surface exceeds that of a smooth or polished one.
3. Loosely laminated materials. This applies particularly to some refractory metals, resin-filled fabric electrical insulation and some natural flaky materials, such as mica.
4. High vapour pressure constituents, in general, for high vacuum work

where baking is to be carried out for the removal of water vapour and other absorbed gases. The following constituents are to be avoided: lead, cadmium, mercury, zinc, sulphur, etc.

5. Low melting point materials. Obviously, if vacuum vessels are to be baked during exhaust, solders or other materials which would weaken, melt or evaporate at the baking temperature should not be used.
6. Materials with high outgassing rates, e.g. most plastics.
7. Permeable materials, e.g. thin rubbers and plastic components.

Materials for vacuum use can be grouped as follows:

1. Metals and alloys
2. Plastics and elastomers
3. Ceramics and glasses

Metals and alloys

Stainless steel   Stainless steel is a high-strength material. It stands up to the wide temperature changes that can be experienced in vacuum work. Also, it does not oxidize, so its surface remains relatively smooth. This means that it cannot produce large oxide surfaces for gas to stick to. Niobium-stabilized stainless steel is preferred. It is easily joined by welding or brazing.

Aluminium   Pure aluminium is best used for argon arc welding but is often brazed and soft soldered successfully. The aluminium–magnesium alloys are the most commonly used. Alloys with a high zinc content should not be used. Corrosion resistance is good.

Nickel   Nickel and nickel alloys are excellent materials for vacuum use. Sheet is useful for pressings and spinnings. Corrosion resistance is excellent.

Copper   High-conductivity copper and oxygen-free high-conductivity (OFHC) copper are widely used in vacuum work. Corrosion resistance is generally good. It is easily machined and joined. OFHC copper is ideal for gasket material.

Brass   This is suitable for some vacuum applications. Brass for flanges should be of plate material, as this is least likely to be porous. It is easily machined and its corrosion resistance is good. However, it is not acceptable at high temperatures since zinc evaporates out.

Mild steel   Since mild steel will rust, it is not usually recommended, but may be used for vacuum systems down to 10^{-5} mbar. Its range may be extended to lower pressures if the surfaces are plated, which should be applied after brazing or welding.

Form of metal   Do not use slices from bar materials for flanges, etc., as bar often contains longitudinal porous 'streaks', which provide leak paths. The orientation of the grain of a material is determined by the direction of extrusion of the bar or the rolling direction of the plate. It should always be parallel to the vacuum wall to minimize penetration of the wall. Air castings, particularly those in brass or iron, are likely to be porous. Vacuum castings are recommended. Forged blanks are particularly suitable for items such as flanges because of the favourable grain direction on both flat and cylindrical surfaces.

Plastics and elastomers
Plastics in general evolve large quantities of gas, and have a high permeability rate compared with other solids. Thus their use in vacuum should be minimized. PTFE has a low outgassing rate and is a good electrical insulator. It can be used at higher temperatures than most other plastics and has good self-lubricating properties.

By far the main use of elastomers in vacuum practice is as gaskets in demountable seals and as valve seals with subsidiary use as tubing for flexible backing lines. As seals their ability to spring back to their original shape makes them reusable in most cases.

Nitrile rubber has a low outgassing rate, good deformation resistance and is oil resistant. Fluoroelastomer has a very low outgassing rate, particularly after baking, and possesses good chemical stability at temperatures up to 200 °C and is also oil resistant. These elastomers are available in a wide range of 'O' ring extruded cord, in sheet and in the case of nitrile rubber in several sizes of thick wall vacuum tubing.

Ceramics and glasses
Ceramic is an excellent insulator. It has good electrical insulation properties, and it also serves to reduce the transfer of heat between components. The fully vitrified ceramics are recommended for use as insulators. Non-vitrified ceramics may be used but must be handled with more care since they readily pick up dirt, which is difficult to remove. Corrosion resistance of all ceramics is excellent. They have limited resistance to mechanical shock.

Glass is a good material for vacuum systems. Borosilicate glass (e.g. pyrex) can be used to assemble complete systems using standard components; it can also be used for viewing windows at temperatures up to 180 °C. With increasing temperature the permeability to gases increases, particularly to helium. Machinable glass-ceramic is another extremely useful material. Note that Kovar is an intermediate material often used in glass-to-metal and ceramic-to-metal joints in vacuum work. Kovar has a thermal expansion rate close to that of glass and ceramic. It therefore prevents distortion and cracking of the joints when the temperature changes.

Fabrication methods

The main effort in this direction is aimed at the avoidance of trapped volumes and crevices, which can slowly release gas when under vacuum and constitute what are known as virtual leaks. Care should be taken at the design stage to exclude such sources.

Joints should be made by screw fixings, or soldering, brazing and welding. Recommended methods of making permanent vacuum-tight joints are in order of preference:

1. Welding—usually arc welding in inert gas atmosphere, i.e. argon arc
2. Brazing—preferably vacuum brazing
3. Soldering—using low vapour pressure alloy

Riveting, staking, etc., are not approved methods of jointing in vacuum because of the likelihood of trapped volumes in joints and the impossibility of disassembly for cleaning.

The number of joints in a vacuum system must be kept to a minimum so that possible leak paths and trapped volumes are avoided. Other considerations regarding the construction of permanent joints are:

1. Bending of pipes is preferred to proprietary pipe fittings, e.g. bends, elbows, etc. This reduces the number of joints and allows for greater ease of cleaning.
2. Never hard-solder proprietary fittings that contain soft solder before removing all traces of soft solder—preferably by cutting off behind the solder groove.
3. Hard solders containing cadmium must not be used.
4. Before welding or soldering valves or gauges, etc., into position, always remove parts liable to be damaged by heat.

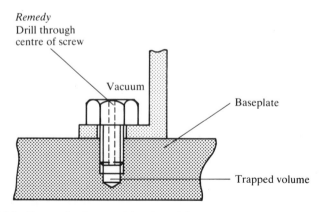

Figure 12.25 Trapped volume arising from blind hole in baseplate

5. Always inspect newly made joints to ensure continuity of the filler material, preferably before they cool down.

An appreciable source of trapped volumes arises from screwed fixings, particularly when tapped blind holes in baseplates are used. An example is shown in Figure 12.25 where a screw holds a bracket in place. A solution is to drill a hole down the centre of the screw, to provide a leakage path for gas trapped at the base of the tapped hole. Alternatively a slot can be milled down the length of the screw.

Figure 12.26 shows an incorrect method of welding a tube to a flange to create a trapped volume. A crack in the inner weld could lead to a slow release of trapped gas during pump-down. Additionally, it is impossible to leak-test the inner weld from the outside if the outer weld is sound. The correct method would be to have a discontinuous outer weld or no outer weld at all.

Screws, bolts, nuts and washers which are cadmium plated should be prohibited from use inside the vacuum system under any conditions and great care must be taken not to confuse them with nickel-plated items. It is doubly important that they do not get into bakable equipment as the high

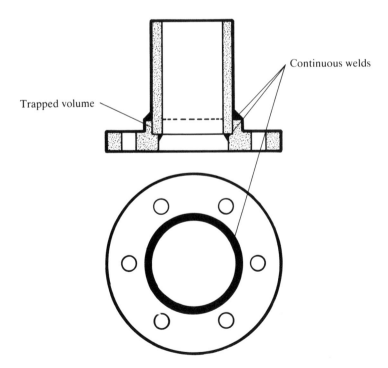

Figure 12.26 Incorrect method of welding a tube to a flange to create a trapped volume

vapour pressure and low melting point of cadmium results in its diffusion throughout and consequent contamination of the vacuum system at quite moderate temperatures.

Parts should fit together without undue strain and matching holes should be in good alignment before tightening the fixing bolts together, as lack of attention to this can result in strained joints leading to subsequent leakage.

When tightening flanges with peripheral fixing bolts and metal seals, tighten gradually in stages bringing all nuts to finger tightness and then tighten diametrically opposite nuts in sequence in easy stages so that no part of the seal is unduly distorted.

Cleaning

Items for assembly into a vacuum environment must be cleaned to remove materials such as flux, greases, cutting fluids, rust-prevention coatings, polishing compounds, etc. (This will have been done on components supplied by vacuum equipment manufacturers.) Such materials will give off vapours under vacuum and retard or prevent the desired low ultimate pressure being reached. The degree of cleaning depends on the pressure level required.

At all stages of the cleaning processes, the work should be handled using an appropriate type of glove. During final stages, lint-free gloves must always be worn as fingerprints on the surfaces of the component can nullify most of the effort used to clean it.

General purpose cleaning to be discussed here is suitable down to 10^{-7} mbar. Procedures for cleaning components to be used in UHV systems is covered in Section 13.6. The golden rule of cleaning is to take care not to leave any remains of the cleaning agent used.

> Do not replace the original source of contamination with contaminants from the cleaning agent used.

For general purpose cleaning the following suggested sequence could be followed:

Mechanical cleaning   e.g. shotblasting
   ↓
Degreasing   e.g. trichloroethane
   ↓
Chemical   e.g. acid
   ↓
Final cleaning   e.g. isopropyl alcohol
   ↓
Drying

Mechanical cleaning
This involves the removal of gross deposits of oxides, welding deposits, etc., with wire brushes, files, grinding wheels or by shotblasting. Shotblasting is suitable for most materials (not plastics). The shot material (glass beads, sand, nickel beads), however, may be difficult to subsequently remove. Obviously during this process sealing surfaces should be protected. Hand-polishing methods include polishing powders, such as jeweller's rouge; steel wool; emery/carborundum and non-impregnated scouring pads. With hand-polishing methods, care must be taken to remove cleaning agents, e.g. abrasives, buffing soaps, etc.

Degreasing
Where possible a vapour degreasing bath with hot trichloroethane should be used. When this is not possible surfaces may be swabbed with or immersed in the cold solvent. Care should be taken to see that the vapour is not inhaled and that unprotected skin is not exposed to the solvent. Where tapped holes, crevices and inaccessible parts are involved ultrasonic cleaning in a solvent bath should be used.

Chemical cleaning

Acids and alkalis　　It must be recognized that chemical cleaning is often hazardous to the component and certain metals require specialized treatment; therefore expert advice should be obtained. However, on occasions components may be pickled in acid or other solutions to remove scale or oxide layers, but since this is an etching process metal will be removed from the surfaces. In general, whatever method is used, it must be followed by rigorous washing and drying to make sure that all traces of the chemicals are removed.

Electropolishing　　This is very suitable for stainless steel (the electrolyte is normally phosphoric acid). However, it is a specialized process and remarks about final washing and drying in the preceding paragraph also apply here.

Final cleaning
Wash with demineralized water (where appropriate) and remove traces of water with acetone or isopropyl alcohol. Dry by warming to perhaps 80 °C until all solvent has gone.

　Note that unless properly dried leaks may be temporarily blocked by solvent; e.g. solvent in capillary channels may take many days at ambient temperature to evaporate. This can result in components being leak-tight at the time of testing (after final cleaning), only to start leaking some time later in service.

Test for cleanliness
The recognized test for satisfactory cleanliness for vacuum purposes, down to UHV pressures, is as follows. Any surface, crevice, joint, weld seam or hole, when rubbed with a clean cotton-wool swab (surgical quality) damped with isopropyl alcohol, should not produce any discolouration of the swab. Note that the swab when applied to the surface should not be saturated with alcohol or this will result in the dissolving and/or distribution of contaminants (if present) over a large area of the swab, making detection difficult.

Handling

Surfaces of components that will be exposed to vacuum should be handled as little as possible after final cleaning. In the case of very clean systems and ultra-high vacuum equipment, the assembler should wear clean cotton, linen or Terylene gloves, which should be changed regularly when soiled.

Surfaces of benches, storage racks, etc., on which components are placed prior to fitting should be kept free from swarf and dust. Components such as high vacuum pumps, liquid nitrogen traps, etc., waiting for assembly to the system should have their openings covered.

Care should be taken in handling heavy baseplates and flanged components not to damage or impair the surface finish of sealing faces. Particular attention should be given to avoiding radial scratches across sealing surfaces. Concentric scratches or scratches following the line of the seal are a lesser problem, especially with elastomer seals, and are sometimes tolerable depending on their depth and extent.

A particular practice to be avoided is sliding one flange on another. Mating components should be brought directly into alignment without any appreciable sliding action. Materials and components should be kept covered when not being worked upon, by polythene or PVC sheets.

13

Considerations in system design

13.1 Conductance

The resistance to gas flow of components that make up a vacuum system (e.g. interconnecting pipelines, valves, baffles, etc.) has a considerable influence on the effective pumping speed and pressure obtainable within the system. In vacuum work it is often more convenient to work in terms of conductance (C) rather than resistance (R):

$$C = \frac{1}{R} \qquad (13.1)$$

In Section 2.8 we saw that the pumping speed (volume flow rate) at any point in a vacuum system can be expressed as

$$S = \frac{Q}{P} \qquad (13.2)$$

where Q is the throughput and P is the pressure of the gas at the point at which the pumping speed is defined.

Therefore, in the case of a pipe where gas is flowing at a rate Q from a region where the pressure is P_1 to a region where the pressure is P_2 then the pumping speeds at the two points are given by

$$S_1 = \frac{Q}{P_1} \quad \text{and} \quad S_2 = \frac{Q}{P_2} \qquad (13.3)$$

The conductance between two points in a vacuum system can be expressed as the quantity flow rate of gas flowing through a device divided by the resulting pressure drop, i.e.

$$C = \frac{Q}{P_1 - P_2} \qquad (13.4)$$

(Compare this with the analogy of current flowing through an electrical resistor.) Substituting for P_1 and P_2 in (13.4) gives

$$C = \frac{Q}{Q/S_1 - Q/S_2} \qquad (13.5)$$

Dividing by Q and rearranging

$$\frac{1}{C} = \frac{1}{S_1} - \frac{1}{S_2} \qquad (13.6)$$

The units of conductance are those of volume per unit time.

If a rotary pump of speed S_p is connected to a system by a pipe of conductance C, the effective pumping speed S_e is related to the rated pumping speed by

$$\frac{1}{S_e} = \frac{1}{C} + \frac{1}{S_p} \qquad (13.7)$$

or

$$S_e = \frac{S_p \times C}{S_p + C} \qquad (13.8)$$

To establish the effective speed of a pump connected to a chamber by a single pumping line at a certain average pressure, the conductance of the pipeline at that pressure must be determined.

The conductance varies as the mode of flow changes and account of this must be taken into consideration when calculating conductance values.

13.2 Gas flow regions

In the pressure region 1013 to 1 mbar the flow is termed *viscous*. Here the gas molecules collide with one another frequently. The mean free path of the molecules is much smaller than the diameter of the pipeline or tube through which the gas may be flowing. Close to atmospheric pressure the flow is in commotion and is known as *turbulent flow*; nearer 12 mbar it becomes layered—generally termed *laminar flow* (see Figure 13.1).

In the pressure region below 10^{-3} mbar the flow is *molecular*. Here the molecules move freely without mutual hindrance and collisions are mainly with the tube wall. Molecules strike the wall, stick for a while, then leave in a new, unpredictable direction. Flow is truly random and the mean free path is much greater than the tube dimension. Pumping occurs only when molecules migrate into the pump of their own accord.

In the transition from viscous to molecular (1 to 10^{-3} mbar) the flow is termed *intermediate*, and here the mean free path is approximately equal to the tube diameter.

13.3 Conductance of pipelines

Figure 13.2 gives the conductance of round smooth-bore straight pipes for air at 20 °C over the range from viscous to molecular flow conditions. The curves in Figure 13.2 are based on the following formulae:

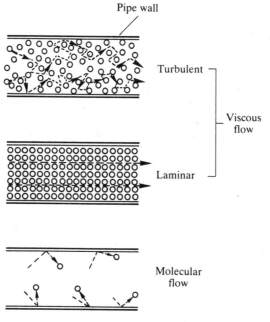

Figure 13.1 Pictorial analogy of the different flow regions

$$\text{Viscous conductance } C_v = \frac{136.5D^4P}{L} \tag{13.9}$$

$$\text{Intermediate flow } C_t = \frac{D^3}{L}\left[136.5DP + 12.1\left\{\frac{1 + 192DP}{1 + 237DP}\right\}\right] \tag{13.10}$$

$$\text{Molecular conductance } C_m = \frac{12.1D^3}{L} \tag{13.11}$$

where D = pipe bore in cm (assuming circular cross-section and straight)
P = average pressure in the pipe in mbar, i.e. $(P_1 + P_2)/2$
P_1 = pressure at inlet
P_2 = pressure at outlet
L = pipe length in cm
C = conductance in $l\,s^{-1}$

The curves can be used to estimate pipeline conductances. For example, to find the conductance of a 5 mm diameter pipe, 2 m long at an average pressure of 10 mbar, the graph indicates a conductance of about $0.8\,l\,s^{-1}$ for a pipe 1 m long. The conductance of a pipe 2 m long is therefore approximately

$$0.4\,l\,s^{-1} \quad \text{or} \quad 1.4\,m^3\,h^{-1}$$

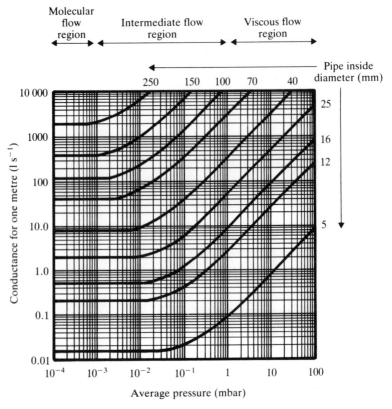

Figure 13.2 Conductance of round pipes for air at 20 °C (litres per second for one metre length)

If this pipe (ignoring any bends) were connected to the inlet of a rotary pump of speed 18 m³ h⁻¹ the effective speed at the end of the pipe, applying Eq. (13.8), is 1.3 m³ h⁻¹, i.e. the pipe has such a low value of conductance compared with the rotary pump speed that it therefore has a very resistive effect and is an overriding factor.

Figure 13.3 shows wrong and right ways of connecting system components. There is a general rule to follow when connecting components.

Connecting pipes should be of short length and large bore.

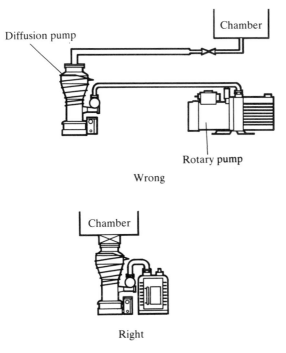

Figure 13.3 Wrong and right ways of connecting system components

Bends in pipes

To take account of short radius elbows when using the formula to estimate the conductance of pipelines, the addition of a length equal to the internal diameter of the pipe is an approximate equivalent. For more gradual bends of large radius, use the length of the tube along its centre-line. These statements apply to all gas flow regions.

Pipes in series (see Figure 13.4)

Overall conductance of pipes or other impedances to gas flow connected in series can be estimated from the following formula:

$$\frac{1}{C}\,(\text{overall}) = \frac{1}{C_1} + \frac{1}{C_2} + \frac{1}{C_3}, \quad \text{etc.} \quad (13.12)$$

where C_1 is the conductance of the first component, C_2 is the conductance of the second component, etc.

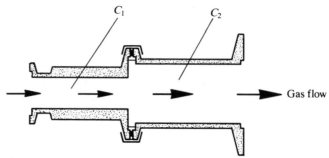

Figure 13.4 Pipes in series

Pipes in parallel (see Figure 13.5)

The overall conductance of pipes in parallel is the sum of each individual pipe conductance, i.e.

$$C \text{ (overall)} = C_1 + C_2 + C_3, \quad \text{etc.} \tag{13.13}$$

Effect of an orifice (see Figure 13.6)

Where gas flow is from a large bore to a smaller bore pipe, there is a further conductance to be considered. This is the transition between the two sections, referred to as the conductance of an orifice. For this situation the following formulae apply:

$$\text{Viscous conductance } C = 20A \quad (\text{l s}^{-1}) \tag{13.14}$$

$$\text{Molecular conductance } C = 11.6A \quad (\text{l s}^{-1}) \tag{13.15}$$

where A = area of the orifice in cm^2

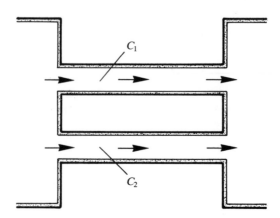

Figure 13.5 Pipes in parallel

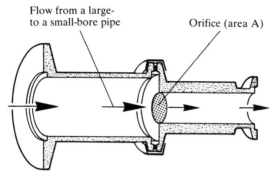

Figure 13.6   Effect of an orifice

An example of the effect of an orifice is illustrated in Figure 13.7. In the left-hand diagram, pump A is shown to be connected directly to a chamber via a $1\,cm^2$ opening. Pump A has a speed of $300\,l\,s^{-1}$ (molecular flow). In the right-hand diagram, pump A has been replaced by pump B which has a speed of $30\,000\,l\,s^{-1}$; the orifice size and chamber remain unchanged. One might be misled into thinking that pump B would give a significantly higher effective speed in the chamber. Actually for *pump A* the effective speed (using Eq. (13.8)) is

$$S_e = \frac{300 \times 11.6}{300 + 11.6} = 11.2\,l\,s^{-1}$$

and for *pump B*

$$S_e = \frac{30\,000 \times 11.6}{30\,000 + 11.6} = 11.6\,l\,s^{-1}$$

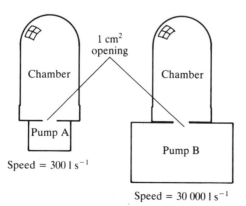

Figure 13.7   A $1\,cm^2$ opening between the chamber and two different sized pumps

Hence, although pump B has a speed that is 100 times faster than pump A, pump B does not give a significantly higher effective pumping speed as might be expected. To increase the effective pumping speed in the chamber requires a larger opening.

The formulae used so far have assumed that air is flowing through the component. For other gases, the result for air is multiplied by a gas correction factor. Some factors are given in Table 13.1.

Table 13.1 Factors at 20 °C

Gas	Molecular flow	Viscous flow
Air	1.00	1.00
Oxygen	0.947	0.91
Nitrogen	1.013	1.05
Hydrogen	3.77	2.07
Carbon dioxide	0.808	1.26
Water vapour	1.263	1.73

The conductance value of pipelines and apertures depends on many factors:

1. Pressure region
2. Cross-sectional shape of tube, e.g. circular or rectangular, etc.
3. Whether straight or bent
4. Molecular weight of gas
5. Temperature of gas
6. Length of duct
7. Surface finish
8. Diameter

13.4 Conductance effect of other vacuum components

The effect on pumping speed of vacuum components such as baffles, traps and valves can be calculated in a similar way to that used for pipelines. The conductance of such components is normally measured practically and quoted by the manufacturer.

Example

A baffle of conductance (C_b) $60\,l\,s^{-1}$ is placed above a diffusion pump of speed $150\,l\,s^{-1}$ and a liquid nitrogen trap of conductance (C_t) $105\,l\,s^{-1}$ is situated on top of the baffle. The overall conductance of the baffle and trap is

$$\frac{1}{C_{overall}} = \frac{1}{C_{baffle}} + \frac{1}{C_{trap}} \qquad \text{[from Eq. (13.12)]}$$

The effective speed (S_{eff}) of this combination at the inlet to the trap is given by

$$\frac{1}{S_{eff}} = \frac{1}{S_p} + \frac{1}{C_{overall}} \qquad \text{[from Eq. (13.7)]}$$

Thus S_{eff} is approximately $30 \, l \, s^{-1}$. The effect of the baffle and trap is to cut the effective speed of the diffusion pump down to a fifth of its quoted value.

Further, if this combination of components is directly connected to a vacuum chamber and the total throughput (Q), in terms of leaks, outgassing, etc., of the system is $3 \times 10^{-6} \, mbar \, l \, s^{-1}$ then the ultimate pressure obtained is

$$P_{ult} = \frac{Q}{S_{eff}}$$

$$= 1 \times 10^{-7} \, mbar$$

If a base pressure of 10^{-10} mbar is required it is not practical to reduce the pressure by increasing the pump speed. In this case, this would require an increase of effective speed by $\times 1000$. The solution would be to reduce the gas load (throughput).

13.5 Gas and vapour load

Probably the most significant factor in the selection of a pumping system is the total gas and vapour load to be pumped; this is affected by the following (see Figure 13.8):

1. Volume of system, mainly the process vessel or chamber
2. Materials of construction and condition of internal surfaces (this relates particularly to outgassing rates)
3. Leakage into the system through joints, etc.
4. Permeation and diffusion through the vessel walls and seals
5. Outgassing of process material
6. Possibly back-migration and/or back-streaming from the pump

Items 1 and 2 above are usually controlled by process requirements and cost. Leakage (3) is not a serious problem if good jointing practice is observed and acceptable leak rates specified (see Chapter 14). Permeation (4) occurs when a pressure difference exists across the surfaces of a vacuum wall. Gas enters the surface where the pressure is higher and diffuses to the vacuum side where it is evolved (see Section 2.4). It is unlikely to be of major proportions except

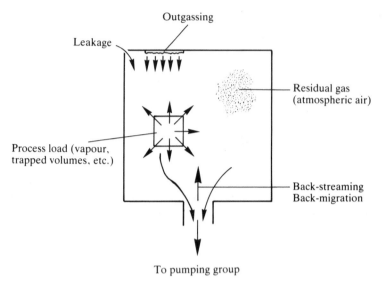

Outgassing

Leakage

Residual gas
(atmospheric air)

Process load (vapour,
trapped volumes, etc.)

Back-streaming
Back-migration

To pumping group

Figure 13.8 Sources of pump gas load

where a large amount of elastomer seal material is exposed to the system, and good vacuum design practice normally eliminates this. Outgassing (5) is frequently the main source of gas load at process pressures and together with item 2 is responsible for most gas and vapour load. With proper vacuum practice using correct accessories and following correct operating procedures, back-migration and back-streaming (6) can be ignored.

The following design criteria relate particularly to item 2:

1. The vapour pressure of any materials used in the construction of a system should be much less than the required ultimate pressure at the operating temperature of the system.
2. The surface of the materials used should produce the minimum of outgassing by being suitably cleaned, preferably polished (to reduce surface area for adsorption) and subject to a minimum of chemical reaction (e.g. rusting) when exposed to the atmosphere.
3. The most common substance to build up many adsorbed layers on a surface is water vapour. This sticks tenaciously to the surface exposed to the atmosphere and its subsequent removal in vacuum can be difficult. It is therefore advantageous to raise a system that is already under vacuum to atmospheric pressure using a gas such as dry nitrogen. Water adsorption is thus eliminated/greatly reduced (provided the system is not exposed to the atmosphere for too long) and during repumping the pressure falls more rapidly.
4. If any appreciably removal of adsorbed gas is required, then the rate of

release is increased by raising the temperature (degassing). Ultra-high vacuum where outgassing is the major source of gas load is seldom obtained without prolonged baking. The stability and vapour pressure of the materials used in the system should be reconsidered to ensure their suitability for bake-out.

5. It is of the utmost importance to design for the maximum effective pumping speed at the vessel.

Outgassing

Regardless of the process through which gas is lodged on or below the material surfaces, when the surfaces are placed under vacuum, gas evolves from these surfaces. The generation of gas by this desorption process is known as outgassing. Outgassing is the spontaneous evolution of gas from a material and becomes an increasingly important proportion of the total gas load once the chamber is roughed down to below 0.1 mbar. Certainly in the high vacuum region outgassing loads must be taken into account when designing systems. For ultra-high vacuum systems, outgassing is the most important factor influencing pump-down time.

The outgassing rate (q) is the quantity of gas given off per unit time by every unit of surface area of a material. Quantity per unit time is the same as throughput and is expressed in millibar-litres per second (mbar l s^{-1}). Outgassing rates are expressed in millibar-litres per second per square centimetre (mbar l s^{-1} cm^{-2}) or pascal-cubic metres per second per square metre (Pa m^3 s^{-1} m^{-2}). Outgassing rates for some commonly used materials of construction are given in Table 13.2.

The following points should be noted:

1. The outgassing rate in vacuum decreases with time as gas is removed from the material. It reaches an approximately constant value after typically four hours.
2. Polymers, elastomers, etc., outgas at rates hundreds of times higher than most metals and glass.
3. Nitrile and fluoroelastomer are two common materials used for 'O' rings. Fluoroelastomer has a lower outgassing rate than nitrile and this fact may be worth considering when choosing the 'O' ring material for a particular application.

An example of how outgassing rates can be used

A process chamber made of unplated mild steel has a surface area of 50 000 cm^2. The seals of the chamber are made of 'Viton A' fluoroelastomer and have a total surface area of 250 cm^2. The chamber is directly connected

Table 13.2 Typical outgassing rates of some common constructional and process materials

Metals and glasses	Outgassing rate ($mbar\ l\ s^{-1}\ cm^{-2} \times 10^{-10}$)	
	1 hour at vacuum	4 hours at vacuum
Aluminium	80	7
Copper (mechanically polished)	47	7
OFHC copper (raw)	266	20
OFHC copper (mechanically polished)	27	3
Mild steel (slightly rusty)	58 520	199
Mild steel, Cr plate (polished)	133	13
Mild steel, Ni plate (polished)	40	4
Mild steel, Al spray coating	798	133
Molybdenum	67	5
Stainless steel (unpolished)	266	20
Stainless steel (electropolished)	66	5
Molybdenum glass	93	5
Pyrex (Corning 7740) (raw)	99	8
Pyrex (Corning 7740) (1 month at atmosphere)	16	3

Polymers, ceramics elastomers, etc.	Outgassing rate ($mbar\ l\ s^{-1}\ cm^{-2} \times 10^{-8}$)	
	1 hour at vacuum	4 hours at vacuum
Celluloid	1 330	665
Kel F	5	3
Mylar (Melinex)	333	66
Nylon	1 596	798
Nitrile	—	63
Plexiglass (Perspex)	400	266
Polyethylene	33	16
Polyurethane	67	33
PTFE (Teflon, Fluon)	40	20
PVC	133	3
Fluoroelastomer (Viton* A)	152	15
Porcelain (glazed)	86	40
Steatite	12	4

* Viton is a registered trade name of Du Pont (UK).

to a pump of speed 280 l s^{-1}. What is the pressure in the vacuum system after four hours (assuming no leakage)?

$$P_{ult} = \frac{Q_{tot}}{S_p}$$

and total gas load

$$Q_{tot} = q_1 A_1 + q_2 A_2 \tag{13.16}$$

where q_1 = outgassing rate of mild steel (at 4 hours)
   A_1 = surface area of mild steel
   q_2 = outgassing rate of 'Viton A' (at 4 hours)
   A_2 = surface area of 'Viton A'

Obtaining values from Table 13.2 and substituting:

$$P_{ult} = \frac{(199 \times 10^{-10} \times 5 \times 10^4) + (15 \times 10^{-8} \times 250)}{280}$$

$$= 3.7 \times 10^{-6} \text{ mbar}$$

13.6 Notes on UHV system design

In Section 13.4 it was seen that it is not generally practical to produce lower and lower pressures by increasing the pumping speed alone. In particular to achieve ultra-high vacuum, among other things, attention must be paid to reducing the outgassing load of materials. There therefore follows a series of notes dealing with UHV system design. The method of design applies to all UHV systems irrespective of what pumps they use.

Reducing outgassing

To help achieve UHV, outgassing can be reduced by the following methods:

1. Eliminate elastomers, hydrocarbon oil and greases.
2. Eliminate materials with potentially bad outgassing properties, e.g. mild steel or porous surfaces.
3. Bake the system; two orders of reduction in outgassing is typical after a 250 °C bake-out under vacuum for a few hours.
4. Eliminate materials that cannot be baked; e.g. brass cannot be baked to high temperatures because zinc is given off.
5. Use clean techniques—gloves, clean atmosphere.

Stainless steel is used for most UHV system materials because it:

1. Has a low outgassing rate.
2. Does not readily corrode.
3. Has an acceptable cost when compared to some alternative materials.

4. Can be readily fabricated.
5. Can be baked up to 750 °C.

Baking

1. Normal bakeout temperatures are 400 to 450 °C for systems exposed to organic contamination and 250 °C for systems free from organic contamination.
2. Maximum baking temperatures may be limited by the maximum temperature that individual system components can be baked to, e.g. seals, valves, glass gauges, magnets.
3. Baking time. System from atmosphere: 20 hours at 350 to 450 °C; subsequent baking of the same system: 2 to 3 hours only at 250 °C.
4. All components in the UHV system must be baked to similar temperatures—otherwise the gas can transfer from a hot surface to a cooler surface and may not be pumped away.

Some examples of maximum baking temperatures are:

	°C
Metal valves with metal pad seal	300–450
Metal valves with Viton 'O' ring on sealing plate	200
Sputter-ion pump with magnets	250
Sputter-ion pump with PTFE lead	250
Sputter-ion pump without lead or magnet	450
Glass ion gauges	400
Fluoroelastomer 'O' rings	200
Wire seals: indium	120–140
gold	400
Copper 'ConFlat®' seal	450

Table 13.3 The effect of changing to metal seals and baking on outgassing rates and ultimate pressure

	Outgassing of walls (mbar l s^{-1})	Outgassing of gasket (mbar l s^{-1})	Chamber ultimate pressure (mbar)
Chamber unbaked fluroelastomer 'O' ring	3.6×10^{-7}	195×10^{-7}	4×10^{-7}
Chamber unbaked ConFlat® gasket	3.6×10^{-7}	0.2×10^{-7}	8×10^{-9}
Chamber baked to 250 °C (with ConFlat®)	1.8×10^{-9}	0.1×10^{-9}	4×10^{-11}

Table 13.3 illustrates how changing to metal seals and baking reduces the system pressure. The data are for a cylindrical stainless steel vessel 30.5 cm diameter $\times$ 30.5 cm high, pumped by a 50 l s^{-1} pump.

To a certain extent the differences between the residual gases left in a baked system compared with an unbaked system will depend on the types of gases evolved from the components of the vacuum system. It will also depend on the type of secondary pump used (see Chapter 11). For one particular UHV system the difference in the residual gas composition before and after baking is shown below (the source of these gases was explained in Chapter 4):

Unbaked system		*Ultimate vacuum* $\simeq 10^{-7}$ mbar
90%	H_2O	
5%	Hydrocarbons (C_nH_m)	
4%	N_2, O_2, H_2	
1%	Inert	

Baked system		*Ultimate vacuum* $\simeq 10^{-10}$ mbar
60%	H_2	
40%	CO	

Wire gaskets

For wire gaskets the flange sealing surfaces need to be lapped/polished with no imperfections such as scratches. High-temperature baking may cause the wire to 'give' (spread), causing a leak. Disc springs should be used under clamping bolt heads to maintain seal loading. Gasket sticking can occur due to intermetallic diffusion between flange and gasket. Other details are shown in Table 13.4.

Table 13.4   Wire gasket thickness and joining methods

Material	Wire diameter (mm)	Compressed thickness (mm)	Joining method
Gold	0.50	0.20	Fusion in bunsen flame; no flux
Indium	0.75	0.13	No prejoining; cold pressure weld when clamping

Surface finish, handling and cleaning

1. Surface finish of materials inside the system should be smooth (but *not* necessarily highly polished). A rough surface has a larger surface area for

gases to stick to than a smooth one and will thus take longer to pump down.

2. Always wear gloves (nylon or terylene) whenever handling components that will be incorporated in the vacuum system and pumped down to UHV. Vapours given off by grease, etc., left by fingerprints will add to the gas load that the pumps have to deal with.

The cleanliness of stainless steel components to be used in UHV systems is of the utmost importance. Components supplied by the manufacturer will already have been cleaned. A suggested procedure for cleaning a component newly fabricated by the user is given below:

1. Vapour degrease with trichloroethane. Immerse in boiling solvent and then lift into the vapour.
2. Rinse with hot alkali (80 °C).
3. Rinse in tap water.
4. Electropolish.
5. Rinse in cold tap water.
6. Rinse in hot deionized water.
7. Rinse with alcohol.
8. Dry with clean, filtered, warm air (60 °C).

The degassing of fabricated components prior to their assembly is useful in reducing the outgassing load on the system. Fabricated components are chemically cleaned to remove surface oxide or other contaminant layers and degreased to remove oil. The component is then heated in a vacuum of 10^{-4} to 10^{-5} mbar to a temperature of the order of 1000 °C. The process is called *vacuum stoving*. (Note that general purpose cleaning of items to be used down to 10^{-7} mbar is dealt with in Section 12.4.)

13.7 Pump-down times and pumping speed—basic calculations

The basic equation for determining the time to evacuate a given volume to a required pressure is

$$T = 2.3 \left(\frac{V}{S}\right) \log_{10} \left(\frac{P_1}{P_2}\right) \tag{13.17}$$

where T^* = time
V = volume of system (litres)
S^* = speed of pump (constant pumping speed is assumed)
P_1 = initial pressure (mbar)
P_2 = ultimate pressure required (mbar)

This formula excludes the effects of outgassing and leakage, i.e. it is for a clean, empty vessel. Take, for example, a chamber of volume 1 m³ which has to be evacuated from atmospheric pressure to 10^{-1} mbar using a 80 m³ h⁻¹

* Time T and speed S must always be in compatible units.

rotary pump. From the above equation,

$$T = 2.3 \left(\frac{1}{80}\right) \log_{10}\left(\frac{1013}{0.1}\right) \quad \text{(time in hours)}$$

$$= 7 \text{ minutes}$$

In a practical case a higher speed pump would probably be chosen to reduce the roughing time. The formula assumes a constant pumping speed over the pressure range considered. Speed curves are not uniform and the speed at 0.1 mbar will be lower than the rated value. One method of dealing with this problem is to calculate the time for each decade of pressure reduced, using an average pumping speed over the decade. The formula simplifies to

$$T = 2.3 \frac{V}{S} \text{ per decade}$$

Figure 13.9 illustrates a simple method of calculating the total pump-down time from the speed curve of a mechanical booster/rotary pump combination. The curve can be approximated to a series of steps as illustrated. The calculation is made in a number of stages utilizing the above formula, which now becomes

$$T_{tot} = 2.3V\left(\frac{1}{S_1} + \frac{1}{S_2} + \frac{1}{S_3} + \frac{1}{S_4} + \frac{1}{S_5}\right)$$

where
S_1 = average pump speed between 1000 and 100 mbar
S_2 = average pump speed between 100 and 10 mbar
S_3 = average pump speed between 10 and 1 mbar
S_4 = average pump speed between 1 and 10^{-1} mbar
S_5 = average pump speed between 10^{-1} and 10^{-2} mbar

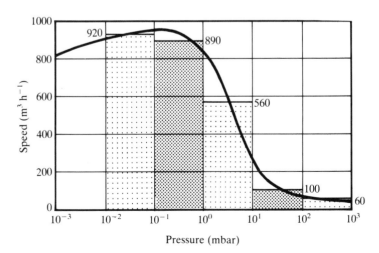

Figure 13.9 Calculation of the pump-down time in stages

Assuming the volume to be evacuated is 10 m³, then

$$T_{tot} = 23 \left(\frac{1}{60} + \frac{1}{100} + \frac{1}{560} + \frac{1}{890} + \frac{1}{920} \right)$$

$$= 0.7 \text{ h (42 min)}$$

Obviously this is a very simplified method and greater accuracy can be obtained by dividing the pumping speed curve into smaller pressure ranges and using the full equation. The above calculations will be further complicated by impedance effects of pipelines and components in cases where the pump is not directly connected to the system.

To allow for impedance effects the effective speed curve at the chamber can be plotted as a function of pressure by taking the conductance of the connecting pipework into account. The value of pipe conductance can be estimated from Figure 13.2 at various pressures and then, using Eq. (13.8), values of effective speed obtained.

An alternative, simple graphical method using the following formula:

$$T = \frac{V}{S} F \qquad (13.18)$$

can be used to estimate the pump-down time from atmosphere, or to determine the size of rotary pump required in the case of a time restraint. A *clean leak-tight* vacuum system directly connected to the pump is assumed. F in Eq. (13.18) is a factor depending on pressure, which can be determined using Figure 13.10.

The following example shows how the graph is used. Assume a pump is required to evacuate a volume of 50 litres to 0.5 mbar in two minutes. The pump speed required is given by

$$S = \frac{F \times V}{T}$$

From Figure 13.10, reading off against the required pressure 0.5 mbar, the pump-down factor F is 8 (for a two-stage pump); therefore

$$S = \frac{8 \times 50}{2} = 200 \text{ l min}^{-1}$$

A pump with a minimum speed of 200 l min⁻¹ (12 m³ h⁻¹) and an ultimate pressure well below 0.5 mbar is required.

Some 'rule-of-thumb' values of pumping speed which have by experience been found to be useful are:

1. For 'gassy' processes carried out in the range 10^{-3} to 10^{-4} mbar, typical pumping speeds of 10 l s⁻¹ per litre of chamber volume are required.
2. For cleaner processes carried out below 10^{-5} mbar a pumping speed of 5 l s⁻¹ per litre of chamber volume is generally adequate.

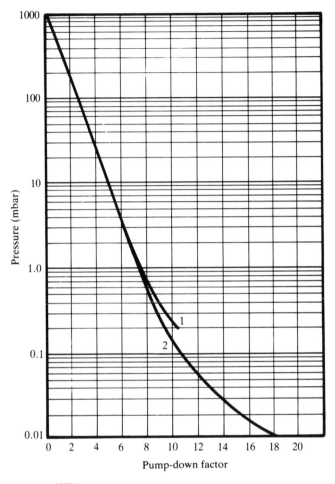

KEY
1 Single stage pump
2 Two stage pump

Figure 13.10 Graph to enable rotary pump size to be evaluated

3. For vacuum processes carried out below 10^{-8} mbar a pumping speed of $1 \, \mathrm{l \, s^{-1}}$ is required for each $100 \, \mathrm{cm^2}$ of process chamber surface area.

13.8 Matching pump combinations

Components are matched by equating throughputs at the inlets to the various pumps and allowing adequate safety factors. However, pump characteristics give rise to some 'rules-of-thumb':

1. *Rotary pump to mechanical booster.* A rotary pump speed of approximately one-eighth the displacement of the mechanical booster is typical.
2. *Mechanical booster to mechanical booster.* The backing mechanical booster has a displacement usually one-third the displacement of the booster working at the lower pressure.
3. *Rotary pump to vapour booster.* The choice of rotary pump is governed by the criterion that the critical backing pressure (CBP) of the vapour booster is not exceeded under conditions of maximum vapour booster throughput (Q_{max}). Maximum vapour booster throughput occurs at about 1 mbar, i.e.

$$S_{rp} \geqslant \frac{Q_{max}}{CBP}$$

A safety factor of 20 per cent is generally included. Thus for a vapour booster with a speed of 200 l s^{-1} at 1 mbar and a critical backing pressure of 3 mbar the minimum rotary pump speed required is

$$S_{rotary} = \frac{200 \times 1}{3} \times \frac{120}{100}$$

$$= 80 \, l \, s^{-1} = 288 \, m^3 \, h^{-1}$$

4. *Rotary pump to diffusion pump.* The rotary pump is matched to a diffusion pump according to the relation given for vapour boosters. Critical backing pressures for diffusion pumps are typically 0.5 mbar and maximum throughput occurs in the constant throughput pumping region at pressures between 10^{-1} and 10^{-3} mbar and can generally be considered to occur at 10^{-2} mbar. Thus for a diffusion pump with a speed of 400 l s^{-1} at 10^{-2} mbar and a critical backing pressure of 0.45 mbar the minimum rotary pump speed required is

$$S_{rotary} = \frac{400 \times 10^{-2}}{0.45} \times \frac{120}{100}$$

$$= 10.7 \, l \, s^{-1} = 38 \, m^3 \, h^{-1}$$

5. *Holding rotary pump to diffusion pump.* Large rotary pumps are required to produce fast roughing cycles and to back large diffusion pumps operating at high throughput. Once these conditions have been fulfilled the large rotary pump can be replaced with a much smaller pump giving lower power consumption and lower noise level. Holding pumps are useful to back large diffusion pumps operating at small throughput, i.e. low process pressures, or operating in an idle condition against a closed isolation valve.

Example

A 36 in diffusion pump with baffle and right-angled isolation valve having a speed at the process chamber of 11 000 l s^{-1} and a critical backing pressure of

0.6 mbar, would normally be backed by a minimum backing pump displacement of 150 m³ h⁻¹ for maximum throughput. A holding rotary pump of 20 m³ h⁻¹ displacement is being used. The maximum diffusion pump throughput that the rotary pump can handle in this situation without stalling the diffusion pump is

$$\frac{20}{3.6} \times 0.6 = 3.3 \text{ mbar l s}^{-1}$$

This throughput is achieved at a process pressure p given by

$$P = \frac{3.3}{11\,000} = 3.0 \times 10^{-4} \text{ mbar}$$

Therefore, the 36 in diffusion pump can be backed by a 20 m³ h⁻¹ holding rotary pump, providing that the process pressure does not rise above 3.0×10^{-4} mbar. Note that in practice when a holding pump is backing a valved-off diffusion pump the rotary pump size is governed by the leakage occurring across the high vacuum valve. This is an unknown factor depending on the state of the valve as a result of being exposed to process conditions, but in most cases it can be anticipated that a holding pump of 10 per cent of the normal backing pump displacement will be adequate.

14

Vacuum leak detection

14.1 The need to control leaks

Much that we take for granted in daily life depends on our ability to control leakage. We drive our cars on tubeless tyres confident that normally wheel rims will lose no air and that fuel tanks will lose no petrol. We expect our refrigerators and deep freezers to operate year after year without any loss of refrigerant from compressors or heat-exchangers. We enjoy television forgetful that television tubes must stay so free of gas that focused beams of electrons inside can reach the screen without any interference to their motion.

Gases like argon, nitrogen and oxygen are transported and stored as very cold liquids in giant steel thermos flasks depending on reliable vacuum (no leakage) between the double walls for thermal insulation.

More and more industrial processes, whether performed under vacuum or not, are performed in plant that must be so soundly constructed that atmospheric gases can be pumped away and any necessary process gases or vapours can be substituted at controlled pressure without the hazard of contamination by leakage in or of pollution by leakage out.

Many essential tools for quality control investigation and fundamental research depend on vacuum of a quality demanding high degrees of leak tightness. Among these are X-ray tubes, mass spectrometers, electron microscopes, surface physics apparatus, space simulation chambers and all sorts of particle accelerator—including the vast machines with which the fundamental structure of matter is investigated and which present most exacting requirements for the control of leakage.

A generation ago, the leak testing of a product was often improvised by the manufacturer—for example by forcing compressed air inside it, immersing it in water or wetting the external surfaces, and looking for bubbles.

Our technological age demands more sophisticated methods. They must be non-contaminating, swift and certain. They must quantify any leakage found so that products can be passed or rejected to a predetermined tightness specification. Suitable methods must be available for a wide range of different products and plant. Leak proving (which passes or rejects a product) and leak finding (which seeks to locate the sites of leakage in a reject) are often required to be separate processes.

14.2 Leak rate

In practice it must be accepted that no container can ever be completely free from leaks. Leakage rates must be kept to a satisfactory low level in order that pumps can effectively produce the required pressure in a system or a product will satisfactorily perform its job over a required period of time.

A leak is any fault in a container wall through which material can pass, from a higher to a lower pressure area.

Leaks are rarely simple round holes; they are usually long, tortuous paths and sometimes include intermediate volumes. In general leaks can be categorized into three general types:

1. *Single gross leaks.* This type of leak might occur, for example, if an 'O' ring has been omitted or severely damaged. Such a leak will at best probably only allow pressure down to a few millibars to be reached. Such a leak is sometimes difficult to locate since they are so large that some leak detection equipment cannot be used. If the background noise level is low it may be possible to hear the leak hissing.
2. *Gross cumulative leaks.* Gross cumulative leaks are usually due to several moderately sized leaks that result in the same symptoms as produced by a single gross leak.
3. *Small single or small multiple leaks.* These leaks may allow the system to pump down into the medium, high or ultra-high vacuum range, but prevent attainment of rated ultimate pressures.

Gas enters the system through a leak at a rate proportional to the external pressure and varies inversely as the square root of the molecular weight of the gas. Since the mol. wt of air is about 29 and helium (the normal trace or search gas) is 4, it is calculated that helium will flow 2.7 times as fast as air through a leak. For most purposes the leak rate can be assumed to be independent of the degree of vacuum and the error involved is negligible.

Let us look at some examples of leakage. One millilitre of air at atmospheric pressure (approximately 1000 mbar) would half fill an average teaspoon. If this amount of air should leak into a vacuum chamber held at a pressure of 10^{-3} mbar it would expand to a volume of 1000 litres. In a chamber held at 10^{-4} mbar its volume would become 10 000 litres and so on. To hold the chamber pressure, the pumps would need to be large enough to extract these volumes in the *time* taken for the leakage to occur.

When discussing leak rates, it is convenient to refer to a quantity of gas passing through the leak in terms of the product of its volume and pressure ($V \times p$). Thus a millilitre of gas at 1000 mbar corresponds to $1/1000 \times 1000$ or 1 mbar l of gas. Similarly 1000 l at 10^{-3} mbar corresponds to 1 mbar l and 10 000 l at 10^{-4} also corresponds to 1 mbar l. This quantity of gas passes through the leak in a time (t).

Thus leak rate is ($p \times V$)/t and the units are normally in millibar-litres per second (mbar l s^{-1}).

Leak rate can be measured in terms of millibar-litre leakage in a second (i.e. in throughput units).

Some other alternative equivalent leak rate units are listed below (see also conversion table 2.4 on page 30):

$1 \, \text{mbar} \, l \, s^{-1} = 0.75 \, \text{Torr} \, l \, s^{-1}$

$= 0.987 \, \text{atm} \, cm^3 \, s^{-1}$

$= 2.097 \times 10^{-3} \, \text{atm} \, ft^3 \, min^{-1}$

$= 1.58 \times 10^5 \, \text{grams per year (Freon 12)}$

$= 7.5 \times 10^2 \, \text{lusec (litre-micrometre of Hg per second)}$

If a process plant is specified to have a leak rate of $10^{-4} \, \text{mbar} \, l \, s^{-1}$ then $1 \, \text{mbar} \, l$ of gas ($\frac{1}{2}$ a teaspoonful at atmospheric pressure) would take 10^4 seconds (2 hours 47 minutes) to enter it. With a leak rate set at $10^{-6} \, \text{mbar} \, l \, s^{-1}$ this time would be increased to $11\frac{1}{2}$ days.

Table 14.1 gives typical leak rate specifications for some vacuum components and apparatus. In general, large plant and apparatus are allowed higher leakage rates than small. It is uneconomic to insist on a degree of leak tightness greater than that needed for the proper working of the apparatus being tested.

For process chambers, valves, pipelines, pumps, etc., to be used in continuously pumped vacuum systems an acceptable leak rate would be such that not more than a tenth of the pumping capacity is needed to pump away all the leakage at the working pressure of the system. This leaves most of the pumping capacity free to deal with the gases continuously desorbing from internal system surfaces and with gases arising from processes performed in the system.

Table 14.1 Typical leak rate specification for some apparatus and components

Item	Leak tightness better than (mbar l s^{-1})
Systems	
Ultra-high vacuum systems	10^{-10}–10^{-9}
'Clean' coating plant and like systems	10^{-5}–10^{-4}
Industrial coating plant and like systems	10^{-4}–10^{-3}
Components	
Ultra-high vacuum pumps and system components	10^{-11}–10^{-10}
High vacuum pumps, valves (incl. seat leakage) and system components	10^{-9}–10^{-8}
Diaphragm valves, backing side components	10^{-6}–10^{-5}
Rotary motion feedthroughs for process plant	10^{-5}–10^{-4}

Leakage entering an apparatus with a working pressure of, say, 10^{-5} mbar produces ten times the volume of gas in the apparatus as would the same amount of leakage entering an apparatus working at a pressure of 10^{-4} mbar—and so needs ten times the pumping capacity to pump it away. It follows that working pressures should not be set lower than necessary and apparatus requiring low working pressures must have leakage rates set at lower rates than those where working pressures are not so low.

For apparatus that must be sealed off, either under vacuum or filled with a working fluid which must not be lost or contaminated by in-leakage (e.g. a bellows thermostat), leak tightness must be specified with regard to the working life expected of the apparatus. Very exacting standards may be required.

There is a miscellaneous group of objects where modern 'tracer gas' methods of leak test have been found effective but where leak rates measured under test do not directly relate to the sort of leakage that might be experienced in final use. Examples are petrol tanks, beer casks and heat-exchangers to carry liquids. In such cases test specifications are derived from practical experience.

An indication of the approximate physical size of a leak is given in Figure 14.1. Here the equivalent hole sizes for different leak rates through a chamber wall of 2 mm thickness are given (assuming a cylindrical capillary hole).

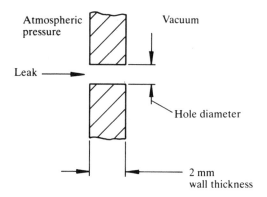

Leak rate $= 10^{-5}$ mbar l s^{-1}	Leak rate $= 10^{-10}$ mbar l s^{-1}
Diameter $= 6 \times 10^{-3}$ mm $= 0.000\ 24$ inch	Diameter $= 1.3 \times 10^{-4}$ mm $= 0.000\ 005$ inch

Figure 14.1   Dimensions of a leak

14.3 The nature and avoidance of leaks

Components required to be leak tight are usually made of metal or glass or ceramic or a combination of these materials. Permanent joints are made by welding, brazing or soldering or the 'wetting' of metal by molten glass. Demountable joints employ gaskets of rubber or metal. Where movement has to be transmitted through the wall of the apparatus a metal bellows may be used or a smooth shaft sealed at the point of entry by a rubber seal lubricated to permit leak-free sliding or rotation.

Leaks in such components are often caused by improperly tightened flanges, 'O' rings that have aged and taken a 'set', improperly seated but undamaged 'O' rings, loose feedthroughs, tiny cracks in ionization gauges, poor fitting or seating of the chamber door gaskets, etc.

Glasses, ceramics and sheet metals do not usually exhibit porosity, but blanks cut from bar can present a problem, particularly near the bar axis, while cast metals are particularly prone to it. Porosity apart, most leaks occur where permanent or demountable joints have been made.

The easiest leaks to discover and locate are those where a definite and quite short path penetrates the wall. Examples are a pinhole where a glass joint is incompletely fused or a metal weld imperfectly made, or a scratched flange against a rubber gasket or a hair on the flange face. In such cases, a tracer gas applied outside the evacuated apparatus can quickly traverse the leak path and appear inside, where its presence can be sensed.

Occasionally the leak path is long and tortuous, perhaps through a complex weld fault with dirt inclusions or cavities or through a thick porous wall, so there is a time lag for a tracer gas to penetrate.

Leaks due to porosity and other causes can also be heat sensitive so that they only appear if the apparatus is heated in use. Leaks at joints can be sensitive to stress, tending to open if unfavourable stresses arise during use. If there are several leaks, it can be hard to find the small ones until the larger ones have been located and cured. Leaks at shaft seals can vary with the motion or position of the shaft.

To help avoid and reduce leakage problems the following points should be considered when designing, making or assembling vacuum components. Some of these points have already been covered in the section dealing with the fabrication of vacuum components (Section 12.4).

1. Design
   (a) Potential leak areas must be accessible for leak testing.
   (b) Avoid creating trapped volumes.
   (c) Correct welding of flanges (weld on the inside).
2. Manufacture
   (a) Use the correct forms of raw material—forgings, not bar for flanges.
   (b) Correct surface finish for sealing faces.

(c) Cleanliness. Machined components should be cleaned to remove swarf and cutting fluid.

(d) Protect sealing faces during transfer.

3. Assembly

(a) Leak-test components individually before assembly.

(b) Check sealing faces for dirt and scratches during assembly.

Virtual leaks

An apparatus can show symptoms of leakage when there is no real leak at all. For example, a rise in pressure inside an isolated vacuum system can originate from outgassing of materials or perhaps from the release of trapped pockets of gas (e.g. from air at the bottom of a 'blind' bolt hole), which seep out slowly over a long period of time. Trapped volumes can be avoided by good design which makes sure that all cavities are well vented.

However, under such conditions a worker could be misled into thinking there is a leak in the system and could spend a great deal of time and money in searching for what appears to be a leak, when in fact the leak does not exist. The term often used to describe this apparent leak is a 'virtual leak'. The difference between a real and a virtual leak is sometimes very difficult to separate.

A total pressure indicator (i.e. a Pirani gauge) can often be used to help distinguish between the two effects; this will show not only the rise in pressure due to the leakage of gas into the system, but also the rise in pressure due to the evolution of vapour. If the system is evacuated to some predetermined pressure, isolated and the pressure rise against time plotted, it will be seen that the curve shows one of two tendencies (see Figure 14.2). From Figure 14.2a it can be seen that the initial pressure rise is rapid, that this tails off at a certain pressure and rises no more. This is indicative of a system that does not contain a leak, but is nevertheless dirty from a vacuum viewpoint. Figure 14.2b shows the same system but containing a leak. In this instance the pressure rise is still rapid initially, but thereafter continues to rise at a steady state. This can be explained by considering the fact that the air leakage into a system will be at a constant rate provided the system pressure is reasonably below atmospheric pressure. The pressure rise due to the 'virtual leak', which is caused by outgassing of contaminants, will vary considerably with change of pressure: the lower the pressure, the larger the volume of vapour evolved. As the pressure in the system goes up, the outgassing tails off and finally stops when the pressure has risen to a value equal to the equilibrium vapour pressure of the contaminants evolved.

Although the pressure rise method may be a little tedious, it does, however, provide useful information:

1. It tells you whether a leak is present. This is possibly the most important information from the user's point of view, as a considerable amount of

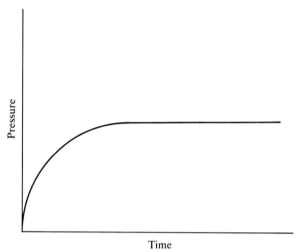

(a) Virtual leak due to evolution of gas or
vapour within a vacuum system

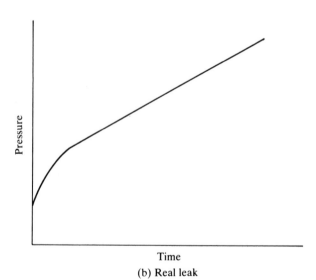

(b) Real leak

Figure 14.2 Distinguishing between a real and a virtual leak

time can be lost in looking for the 'virtual leak'—not to mention the general wear and tear on the nervous condition of the user!
2. It gives a good indication of the condition of the system in terms of its cleanliness.

14.4 Leak detection methods

Having established the fact that a leak is present in the vacuum system, there now remains the choice of method used to pinpoint it. The detection of leaks is often most successfully achieved using (helium) trace gas and a mass spectrometer system. Such methods are typically a million times more sensitive than alternative rough methods.

Helium gas is commonly used as the tracer gas because it is rare in air— only five parts per million in the atmosphere. It is composed of light inert non-toxic atoms which can penetrate small leaks faster than air molecules. Hydrogen is even lighter and faster but is not inert, while argon is heavier, slower and 1 per cent of air. Simple mass spectrometers can easily distinguish helium from other gases and contaminants, so helium is a nearly ideal tracer gas for leak detection.

Leak detection can be accomplished by vacuum testing, where the detector is a part of the vacuum system and the search gas is sprayed over the outside of the component, which is also connected to the vacuum system. If a leak is present the gas is drawn into the system through the leak and the detector will respond. Certain items may also be leak-tested by pressurizing the inside of the object with the search gas and probing the external surface with a sampling probe connected to the detector.

In many cases, because of the construction or nature of the product or system under leak-test, only one of these alternatives can be used. For

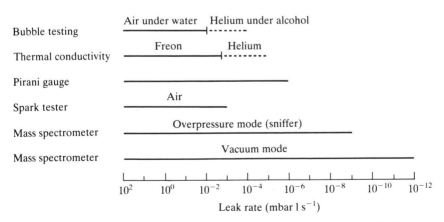

Figure 14.3 Likely sensitivity range of the various leak detection methods considered (using helium unless otherwise stated)

example, 'O' ring seals on vacuum vessels may develop leaks under the reverse differential condition of overpressure testing, while evacuated pressure vessels may collapse under the inward stresses.

Some rough methods of both vacuum and overpressure methods will be described as well as the more sensitive mass spectrometer leak detector method. Figure 14.3 gives an indication of the range of leak sizes that each individual method considered will find.

Masking compounds

Suspected leaks can be covered with a masking compound such as 'Apiezon Q', a low vapour pressure material having the consistency of plasticine. A fall in system pressure will indicate that a fault has been covered. Careful removal piece by piece will pinpoint the leak. Leaks can be masked off to allow further testing; however, it should be regarded as a temporary measure only. The compound applied should be removed and the fault repaired as soon as possible. If left, the compound tends to harden and crack in time, and the leak may once again become evident.

Spark testers

The spark tester readily locates leaks in evacuated glass systems. The tester consists of a high-frequency Tesla coil. The spark discharge from the coil is passed over the glass surface and, provided that the system pressure is in the range 10^{-3} mbar to about 100 mbar any small cracks or pin-holes will cause the spark to concentrate and pass through at that point creating a glow discharge within. Leaving the discharge in this condition for any length of time should be avoided as local heating is created at the fault and the hole may be extended by cracking of the glass. Leaks down to 10^{-3} mbar l s^{-1} may be found.

Bubble testing (overpressure method)

This was widely used for small items and sometimes for plant before more technically advanced alternatives were available. The system or component is pressurized with air and immersed in a clear liquid, e.g. water. The presence of a leak is detected by the ensuing stream of bubbles issuing from it. In this way it is possible to detect leaks of the order of 10^{-2} mbar l s^{-1}.

Some advantage is obtained if the gas is less dense than air and the liquid has a low surface tension. Thus it has been shown that leaks 100 times smaller can be detected with helium bubbles in alcohol than with air bubbles in water.

Obviously the method is somewhat limited by the size of the component under test and the time one is prepared to wait for one bubble to appear (see

Table 14.2   Comparison of the average time for one bubble to form in an immersion test using air and water, with typical leak rate values

Time to form one bubble	Leak rate ($mbar\ l\ s^{-1}$)
10 seconds	2.7×10^{-3}
40 seconds	6.7×10^{-4}
100 seconds	2.7×10^{-4}
400 seconds	6.7×10^{-5}
15 minutes	3.3×10^{-5}
100 minutes	5.3×10^{-6}

Table 14.2). It is also dependent on operator attention span. Sometimes bubbles are from other sources than leaks and give false signals. Another disadvantage is that the object being tested requires drying. A variant of this method, rather than immersion, is to paint or spray the surface to be tested with a soap or detergent solution.

Thermal conductivity detector (overpressure method)

This depends on the differences in thermal conductivity between a search gas and air. The detector (see Figure 14.4) consists of a hand-held unit terminating in a fine sample probe containing a heated filament. An identical filament is housed nearby, as shown, and has a reference probe which terminates at the rear of the unit. Both filaments constitute part of a Wheatstone bridge circuit. A small fan draws in air from the background conditions to supply each filament separately. If a leak is present, in a pressurized component under test, the filament in the sample probe receives

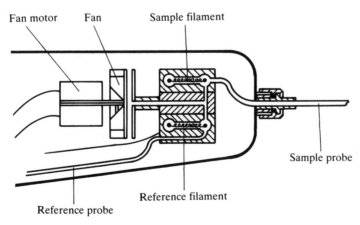

Figure 14.4   Thermal conductivity detector ('Sniffer').

the leaking gas in concentration and, depending on the thermal conductivity of this gas, changes its temperature and consequently its electrical resistance. During this time the reference filament is still sampling the background conditions and therefore does not change in temperature or resistance. The difference in electrical resistance causes an out-of-balance condition which can be read on a meter; audible warning via an amplifier unit can also be provided. Using this unit and helium, leaks down to 10^{-5} mbar l s^{-1} are detectable, and using freon down to 3×10^{-3} mbar l s^{-1}.

Thermal conductivity gauge (Pirani)

We have already seen in Chapter 3 that a Pirani gauge which is calibrated for dry air indicates a different reading when the gauge is filled with helium at the same absolute pressure. Hence, when the atmosphere of air surrounding a vacuum system under test is replaced by an equivalent atmosphere of helium, a Pirani gauge connected to the system will indicate a pressure increase if a leak is present. This type of detector is capable of a sensitivity of 10^{-5} to 10^{-6} mbar l s^{-1} when using a special amplifier unit.

It is customary when leak testing large plant to use a hood or plastic envelope so that sections of the plant may be individually surrounded by search gas and tested. Best sensitivity is obtained at pressures below 0.1 mbar. The Pirani gauge is best located in the backing line.

In vacuum testing, if testing is conducted by spraying helium on the outside of the system locally with a small nozzle, it must be remembered that with any gas that is lighter than air testing should proceed from the top downwards, and conversely with heavier-than-air gases, blocking off leaks as found. Effects of draughts must be minimized.

Mass spectrometer leak detectors

Special mass spectrometers usually of the magnetic deflection type (see Chapter 4) have been developed for use as leak detectors, and these developments have led to the most sensitive of the various leak detection methods available.

Operation of the device depends upon its ability to separate and to measure quantitatively gas ions of a certain mass-to-charge ratio. The detector is normally tuned to detect helium.

Commercial detectors like the fully automatic microprocessor-controlled model shown in Figure 14.5a and b are self-contained vacuum systems; see the vacuum layout schematic in Figure 14.6. The system is connected via the test port to the component to be tested. With the machine illustrated, leaks in the region from 2000 to 2×10^{-12} mbar l s^{-1} can be found. A liquid nitrogen cold trap protects the mass spectrometer analyser head from contamination with condensable gases. If the leak detector is operated with the cold trap unfilled it will not be able to operate at maximum sensitivity because of the higher-than-normal pressure in the analyser. In some applications minimum leak rate testing may not be necessary and no-nitrogen leak detectors are more convenient than top performance versions. No-nitrogen leak detectors that are available utilize either a dual diffusion pump system, a turbomolecular pumped system or a diffusion pumped system and a helium counterflow technique. Considering Figure 14.6, the gas to be sampled is drawn into the system from the test piece. Initially down to pressures of about 0.1 mbar the roughing pump alone is used. With the test valve closed at these high pressures, sampling is achieved through different-sized small 'leaks' which bypass the test valve. For smaller leaks at lower pressures, the roughing valve closes and the test valve automatically opens and allows all the sampling gas to pass into the high-vacuum side of the mass spectrometer. If helium is sprayed over the outside of the component under test and this has a leak, the gas sample drawn into the mass spectrometer will contain helium and its pressure will be registered.

The leak detector should be calibrated and adjusted for maximum sensitivity at frequent intervals. A standard helium leak of the type shown in Figure 14.7 lends itself readily for this purpose. The device consists of a container from which helium escapes at a constant rate by permeation through a quartz membrane. The leak rate is typically 10^{-8} mbar l s^{-1}. This is sufficiently small compared with the amount of gas in the container to ensure nearly constant leak rates for long periods of time. If fitted into the mass spectrometer shown in Figure 14.5 automatic tuning and calibration can be achieved.

Figure 14.8 shows how mass spectrometer leak detectors can be used in a variety of ways to find leaks. In Figure 14.8a the object under test is evacuated. The object is enclosed in an envelope or hood (metal chamber, plastic bag, etc.) which is filled with the tracer gas. The gas enters the vacuum system through any leak paths available. The method does not locate the leak paths.

In Figure 14.8b the object under test is mounted inside a vacuum chamber, the detector being connected to the vacuum system used to evacuate the chamber. Tracer gas at any suitable pressure is supplied *inside* the object under test and enters the vacuum system if there is a leak path available. Again, leak paths are not located.

When permanently sealed objects have to be tested, it is sometimes possible to employ this method by sealing a small amount of tracer gas inside the object (if this does not interfere with its functioning).

In Figure 14.8c a large chamber is being tested in the same sense as the object in Figure 14.8a. The chamber is too large to be enclosed in an envelope so regions where leaks are most probable are being separately 'hooded' for the local application of tracer gas. In the diagram a plastic hood has been taped around a flange joint and a cup is being used to envelop a feedthrough.

(a)

Figure 14.5 (a) A fully automatic microprocessor-controlled helium mass spectrometer leak detector with simple two-button testing. (b) Illustration showing main component layout (opposite)

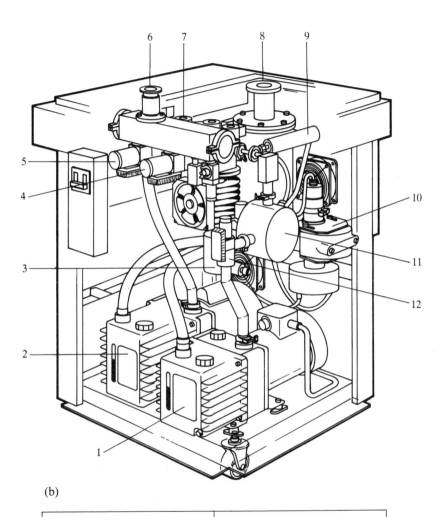

(b)

1	Backing pump	7	Gross leak bypass valve
2	Roughing pump	8	Liquid nitrogen trap filter
3	Air-cooled diffusion pump	9	Quartz reference leak
4	Test valve	10	Mass spectrometer
5	Roughing valve	11	Backing volume
6	Test port	12	Backing valve

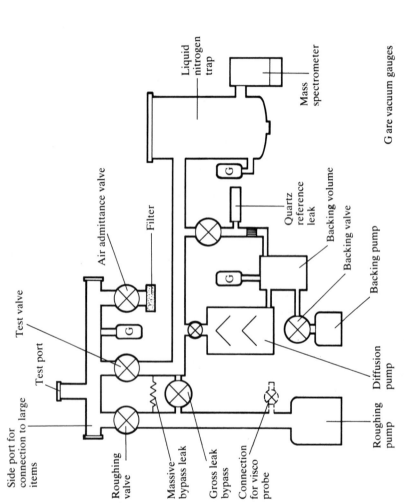

Figure 14.6 Mass spectrometer leak detector–vacuum system schematic

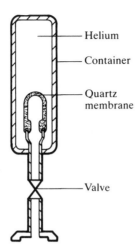

Figure 14.7   Standard helium quartz reference leak

Figure 14.8d illustrates the location of leaks by the use of a fine jet of tracer gas supplied through a hand-held probe nozzle equipped with flow adjustment and a push-button on/off valve. It is being used to find the exact sites of leaks after the methods shown in Figure 14.8a or c have proved an object to be leaky. Likely places are probed with the gas jet and the detector signals when the jet plays on a leak path. The signal can be audible as well as visual. A fine jet can be used to pinpoint a leak site which can be marked for subsequent repair or temporarily sealed so the search for any other leaks can proceed uncomplicated by its presence.

Figure 14.8e illustrates the 'overpressure' use of tracer gas to locate leak sites. The object under test is filled with tracer gas to a safe pressure above atmospheric (proper precautions must be observed) so that some gas flows out through any leakage paths present. Likely leakage sites are sampled or 'sniffed' using a hand-held probe arranged, in this case, to suck in the gases it encounters and deliver them to the detector. Leak sites are pinpointed where concentrations of tracer gas are found. Sniffing commences at the bottom of the test piece to eliminate the possibility of locating a leak when actually the 'wake' of a leak lower down is present.

Figure 14.8f illustrates a method of testing small hermetically sealed objects (e.g. cavity-type integrated circuits, barometer capsules) which have no port available for connecting to a vacuum system or tracer gas supply. It is known as the 'bomb' testing method or 'bombing' method.

The object is placed inside a chamber which can be supplied with tracer gas at high pressure. After a suitable 'soak' period, sufficient tracer gas will have penetrated any leak paths present to build up a concentration inside the object sufficient for test purposes. If now the outer chamber is flushed with,

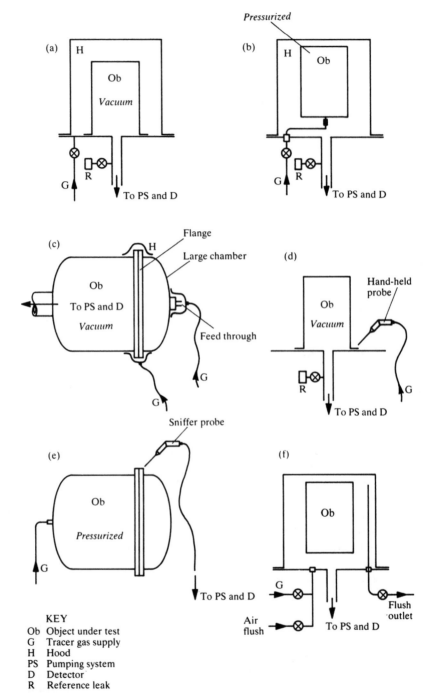

Figure 14.8   Ways of using a mass spectrometer leak detector

say, nitrogen and evacuated by a system containing a detector (or the object is transferred for test to the apparatus shown in Figure 14.8b) tracer gas will emerge through the leak path and indicate the presence of a leak.

14.5 Response times

When testing by a tracer gas method employing vacuum there is a time lapse (usually brief) between the application of the gas upstream of the leak path, its emergence into the vacuum space downstream and its detection.

Concentration builds up more rapidly in a small vacuum space than in a large one. It builds up more rapidly and to a higher final value if the pumping speed used to evacuate the vacuum space is low.

When seeking for leaks with a gas probe a practical time for applying the probe to a particular region is 2 to 10 seconds.

It is important to be methodical when going over a vessel/system with the search gas. When a leak is indicated, go over the area already covered more slowly to establish the exact location of the leak.

When tracer gas is removed from the entrance to a leak path there is a delay before the leak path empties itself of tracer gas and a further time lapse before the tracer gas in the vacuum space has been pumped away to a low enough level to permit the search for further leaks to proceed. Rapid 'clean-up' is favoured by a small-volume vacuum space and by a high pumping speed.

The ratio of chamber volume to pumping speed (V/s) is sometimes called the time constant or response time of the system. A low value of V/s (small volume, high speed) favours rapid response. However, the much greater tracer gas pressure rise achieved by keeping pumping speed small is the dominant consideration for leak testing.

Specialized leak detection plant often employs valving for quick control of pumping speed.

14.6 Some specific examples of the use of mass spectrometer leak detectors

Examples are given below of some of the tasks that mass spectrometer leak test systems can undertake.

Example 1 (see Figure 14.9)

A manifold is used to test a number of batch-produced items simultaneously (vacuum valves in this case). A single hood can be used to cover the whole manifold, filled with helium from beneath with the spray probe. An individual hood can be used to identify the culprits if any leak is present. Valves are first tested open with the free ports blanked and then closed with the free ports open.

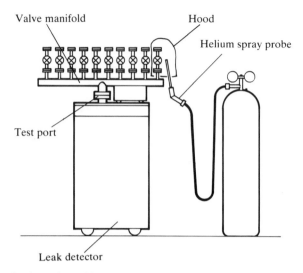

Valve manifold Hood

Helium spray probe

Test port

Leak detector

Figure 14.9 Leak testing of batch-produced items

Example 2 (see Figure 14.10)

The leak detector is provided with a large-area work top for the testing of big items. An extension attached to the end port of the test manifold is used to make the pumping connection. A plastic bag is a convenient hood for all-over testing. If leaks are present, a hand-held spray probe is used to locate them; leaks found can be pinpointed by adjusting the spray probe to have a small jet.

There are sensible limits to the use of the leak detector's pumps alone for very large items. For example, an over-long roughing time may be required. The time constant can be unacceptably high for a very large item; with an item of 100 litre volume and a test port pumping speed of, say, $10\,l\,s^{-1}$ the response time (V/s) is 10 seconds. Therefore, spray probing must be performed very slowly when searching for small leaks. Additionally, it is unwise to use the high vacuum pump (diffusion pump) in the mass spectrometer leak detector repeatedly to evacuate very large volumes; it must labour too long at the high pressure end of its operating range.

Figure 14.10b shows a work station for leak testing large items equipped with a mechanical booster–rotary pump combination pumping set. A pipeline from the leak detector test port is connected to the outlet between the mechanical booster pump and the rotary pump. There is a throttle valve to the rotary pump inlet; this is used to divide the pumping speed between the leak detector and the rotary pump to ensure that an appropriate sensitivity of test can be obtained. The leak detector inlet test valve need not be opened until pump-down is complete. The advantages of sampling tracer gas from the backing side of a fast secondary pump are discussed in Example 3.

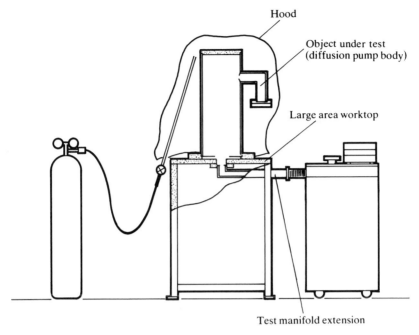

(a) Using the pumps in the leak detector

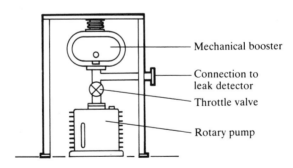

(b) Leak testing work station with mechanical
booster/rotary pump combination

Figure 14.10 Leak testing big items

Example 3 (see Figure 14.11)

The leak detector test port is connected to a purpose-built leak test port on
the backing side of a diffusion pump serving a large vacuum processing plant.
A helium spray probe and a control module extension is being used.

Because the diffusion pump has high speed the time constant for the high
vacuum chamber is likely to be small. In an actual example a 600 l chamber

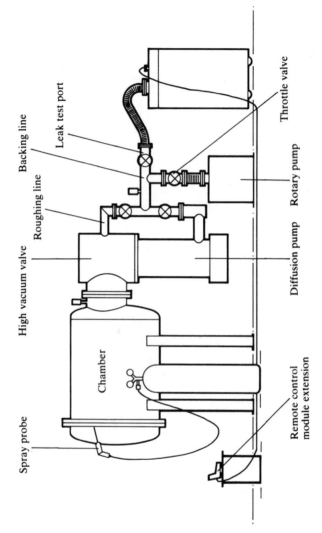

Figure 14.11 Leak testing a large vacuum processing plant by sampling gas in the backing line

was pumped by a 3000 l s^{-1} pump system so the response time (V/s) was 0.2 seconds. This means that only a fraction of a second after helium starts entering through a leak it is being transferred to the backing system.

Backing pipeline volumes in comparison are usually small, giving a short response time for the backing side. In the example the backing volume was about 12 l, with the backing pump speed about 50 l s^{-1} and the response time 0.24 seconds, so the helium concentration in the branch line to the leak detector would approach its maximum value only a fraction of a second after helium started entering through a leak.

All the helium entering a leak is shared between the leak detector's pumping system and the plant's backing pump. If the backing pump pumps three times as fast as the leak detector system then only one part in four of the helium gas goes to the leak detector and its sensitivity is therefore reduced. If needed, extra sensitivity can be had by throttling or valving off the backing pump, provided that the critical backing pressure of the plant's diffusion pump is not exceeded.

Example 4 (see Figure 14.12)

This example illustrates the use of the 'sniffer probe' for overpressure testing of items unsuitable for evacuation. The sniffer probe is connected to the test port of the leak detector by a long length of vacuum tubing (typically 2 m). Gas enters the probe through a capillary restriction so that the pressure in the leak detector is kept at about 5×10^{-5} mbar. The response time is about 2 seconds and leaks down to 5×10^{-9} mbar l s^{-1} can be found and measured.

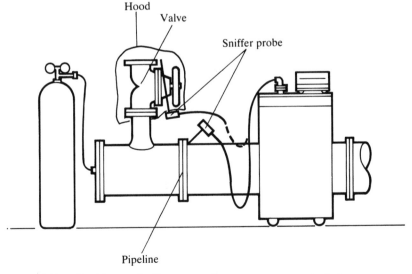

Figure 14.12 Checking a pipeline system by an overpressure technique

The figure shows part of a pipeline system being pressurized with helium gas from a regulated cylinder. The sniffer probe is being used to check flange welds and pipe joints by sucking in any helium emerging from a leak. The valved branch is receiving an all-over test. It has been 'hooded' with a plastic hood and the space under the top of the hood is being probed. Light helium gas would accumulate there if there were any leaks.

When probing for helium with a sniffer it is sensible to start low down and work upwards.

Example 5 (see Figure 14.13)

This example illustrates the use of a device called a 'visco probe' for overpressure testing of items unsuitable for evacuation and where the leak detector itself cannot be brought within two metres of the area to be tested. The visco probe consists of a long flexible pipe (up to 25 m) stored on a reel. The pipe carries a hand-held probe nozzle at one end; the other end is

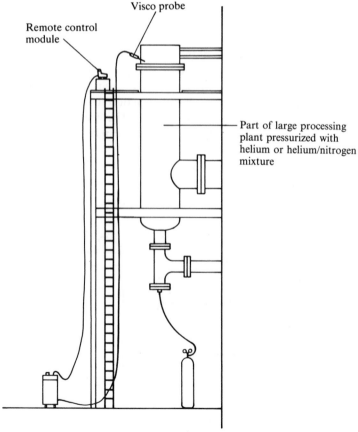

Visco probe

Remote control module

Part of large processing plant pressurized with helium or helium/nitrogen mixture

Figure 14.13   Using a 'visco probe' to overpressure test a large processing plant

connected to a valve just above the leak detector's own roughing pump inlet (see Figure 14.6). Note that the tube length is too great for transfer of tracer gas in the molecular flow mode to be practical; transit times would be too great.

When the visco probe is in use, air is sucked into the probe nozzle and along the pipe in the viscous flow mode, entering the leak detector's roughing duct and through the small leaks bypassing the test valve.

Only a proportion of the helium reaches the mass spectrometer pumping system and sensitivity is limited accordingly. The longer probe gives a response time typically of 7 seconds and a sensitivity of 3.5×10^{-7} mbar l s^{-1}.

Example 6: bomb testing

The 'bombing' method is used to test hermetically sealed devices with small internal cavities which cannot be evacuated or pressurized in the normal way. Sealed pressure-measuring capsules, reed relays and micro chips are examples. Cavity volumes can range from about 0.01 ml up to about 10 ml.

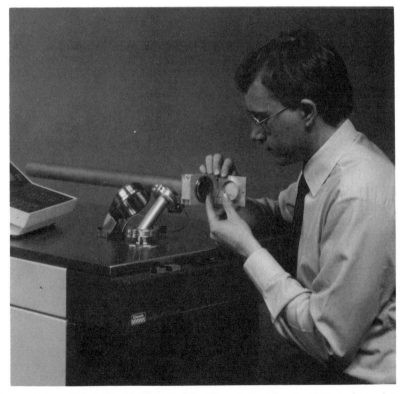

Figure 14.14 'Flip top' and 'stick' test chamber options for the testing of a variety of small sealed devices which have been previously subjected to helium 'bombing'

The method is more sensitive for the smaller cavities. With acceptable test times the method can detect leaks between about 10^{-9} and 10^{-5} mbar l s^{-1} in 0.01 ml cavities and between about 5×10^{-8} and 10^{-2} mbar l s^{-1} in 10 ml cavities.

The bomb testing station shown in Figure 14.14 employs a small vacuum test chamber with options to suit different sorts of test object.

Items to be tested are first placed in a pressure chamber which is flushed with helium and then pressurized to about 4 atm. 'Soaking' at pressure may last between 2 and 4 hours. During this period helium penetrates any leak present and enters the internal cavity where it mixes with the gas already present.

After this pressure treatment, items are allowed to 'dwell' in atmospheric air for a short period (say 10 to 60 minutes) so that superficially adsorbed helium can desorb from the external surface. They are then placed in the vacuum test chamber which is evacuated by the leak detector pumping system. Under vacuum any leak path carries an outward gas flow composed of the original gas content of the cavity mixed with helium which entered during the period under pressure. The helium flow is detected and measured by the leak detector which is able to process the result and to display a total leak rate and a pass/fail indication for the chamber load.

14.7 Dedicated leak detectors for high-volume industrial production

Mass spectrometer leak detectors are increasingly employed by industry to inspect a wide variety of products. Often they are designed to test one specific product only, working automatically or semi-automatically and integrated into the production process where high-volume production is undertaken.

The overpressure method of testing is usually employed. One or more items to be tested are loaded into an individually designed test chamber connected to the vacuum system. The items and the chamber together are partially evacuated; then the items are pressurized internally with helium of helium/nitrogen mixture and the chamber is fully evacuated. There are sensors to indicate the presence of massive leakage; if this is absent the chamber atmosphere is sampled by the mass spectrometer.

There is a wide variety of system arrangements, depending on product, specification and the degree of automation required (stages such as loading, unloading and pressure line connection are sometimes performed manually). Sometimes twin chambers are used so one can be loaded while the other is 'on test'. Typical overall testing times lie between 10 and 40 seconds per item. Helium-recovery systems may be incorporated.

Examples of items tested in this way are beer barrels, gearboxes, refrigerator compressors and heat-exchangers; Section 14.8 contains a fuller list. Test specifications may vary from about 5×10^{-2} to below 10^{-7} mbar l s^{-1}. Figure 14.15 shows a semi-automatic system for testing

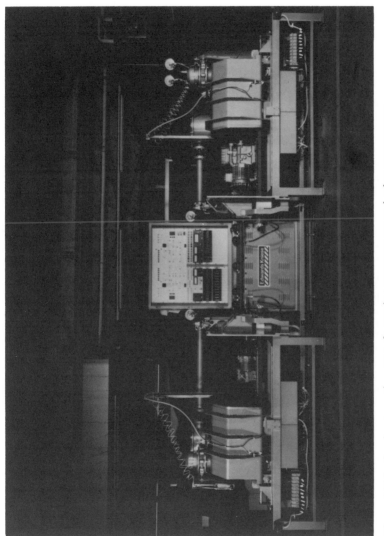

Figure 14.15   Semi-automatic system for testing compressors and other components up to 40 l in volume. The plant is designed to be integrated into a production line

compressors. This is a twin chamber plant with a throughput of 250 to 300 items per hour. Test pressure is adjustable up to 25 mbar and sensitivity is 3.2×10^{-5} mbar $l\, s^{-1}$.

14.8 Applications of helium mass spectrometer leak testing

During the last ten years the applications of modern electronic control and ingenious system design have enormously increased the uses of the helium leak detector. It has become an option worth considering almost wherever there is a need to control leakage, and more often it is found to be the best option available.

Some items it is being successfully employed to test are listed in Table 14.3.

Table 14.3   Items tested by the helium leak detector

Items suitable for testing under vacuum	Items suitable for over-pressure testing using chamber or snifter technique as appropriate	Items suitable for 'bomb' testing
Vacuum plant and components	Process plant	Semiconductor devices
Lamps	Heat-exchangers	Sealed relays and contacts
Electronic valves	Evaporators	Quartz oscillators
Cathode ray tubes	Hydraulic or pneumatic systems	Flash lamp bulbs
Laser tubes	Refrigeration/air-conditioning systems	Sealed diaphragm capsules
Feedthroughs	Hermetic and semi-hermetic compressors	Watch cases
Pacemakers	Pipe runs	Sealed items of small internal volume and able to withstand external pressure
Bellows	Valves	
Pipe assemblies and similar metal fabrications	Filters	
Castings	Thermostats	
Gas valves and regulators	Measuring elements	
Heating elements	Gyroscopes	
Heat-exchangers	Nuclear fuel rods	
Containers of all sorts capable of withstanding atmospheric pressure	Torque converters	
	Gearboxes	
	Shock absorbers	
	Wheels for tubeless tyres	
	Pressure vessels	
	Airships and balloons	
	Petrol tanks	
	Casks and barrels	
	All sorts of container incapable of withstanding external atmospheric pressure	

15

Guide to the safe use of vacuum equipment

In the past many reported incidents and mishaps in the vacuum industry have occurred because the user did not have a thorough understanding of the correct use of the product. This may have occurred through a lack of training about operating procedures or through a failure to comprehend what might happen under certain operating conditions.

A little forethought would have saved the sight of an operator who unfortunately exposed a capsule dial gauge to an extremely high pressure (the glass front ruptured). It would have saved the life of the maintenance worker who was unintentionally trapped in a large vacuum chamber which was then evacuated.

In one reported incident, the oil in a diffusion pump was being removed during maintenance, prior to putting in a new charge of oil. The technique unwisely adopted was to heat the oil at atmospheric pressure in the isolated pump, using the pump's own heater. The objective of heating was apparently to thin the oil to make subsequent draining on release of the drain plug easier. In this case when the drain plug was removed, the hot oil came out, evidently under some pressure and almost at once it ignited. It was found that there had been no control of the time or rate of heating of the oil which must have been at an unnecessarily high temperature for some time.

In another similar incident, a minor explosion (due to ignition of the oil) occurred when a vapour booster was restarted without the atmospheric air first being pumped out of the booster by the rotary pump.

In general, systems to be operated by unskilled personnel should be fully automatic and as foolproof as possible with the number of operations to be carried out by the operator kept to a minimum. With regard to maintenance, correct procedures must be laid down and strictly adhered to.

The following notes are intended as a general guide to some of the precautions necessary for ensuring the safe installation, use and maintenance of some vacuum products.

15.1 General precautions

General safety

1. For safety it is essential that all products are installed, operated and maintained in accordance with the instructions supplied with them.
2. Unless specifically supplied for such use, most products are *not* suitable for use outdoors or in the presence of flammable atmospheres, or in any location for which special safety regulations apply or special precautions are necessary.
3. Vacuum systems do not provide protection against toxic, biologically or chemically hazardous materials that may be processed therein or be evolved as a result of processes that may be carried out under vacuum. It is necessary for the user to provide appropriate safeguards.

Electrical safety

For safe electrical operation of equipment connected to the mains it is essential that the following requirements are met:

1. The electrical installation of the building in which the equipment is installed and all fittings used for mains connection must conform to the Institution of Electrical Engineers Regulations for the Electrical Equipment of Buildings or, if outside the United Kingdom, to the International Electrotechnical Commissions Recommendations.
2. The equipment must be connected to an electrical supply of the correct voltage.
3. The protective earth conductor must be properly connected and good earth continuity maintained at all times.
4. Only fuses and protective devices of the specified ratings and types must be fitted.
5. All electrical equipment must be disconnected from the supply before modifications, repairs or maintenance are undertaken.

Damaged equipment

Equipment that is damaged or is possibly damaged should not be used until properly repaired and its safe operation ensured.

15.2 Hazards associated with particular equipment

Vacuum measuring instruments

General
1. Most vacuum gauges are designed to measure pressures lower than atmospheric pressure and should not be subjected to pressures significantly above atmospheric pressure.

2. Vacuum systems, which may be subjected to high overpressure in the event of malfunction, should incorporate suitable pressure relief devices.

McLeod gauges
1. These gauges are filled with mercury, which is a toxic substance and should be used and handled with great care.
2. The gauges are of glass and to avoid the risk of damage should only be used with the plastic cover that they are normally supplied with in position.

Ionization gauges

Cold and hot cathode types   The heads of these gauges operate at a high voltage. Connection and disconnection of the gauge head from the supply should only be done with the gauge supply disconnected from the mains.

Hot cathode type   These gauges operate with an incandescent filament in the vacuum system. Flammable or explosive gas mixtures must not come into contact with the filament, otherwise an explosion could occur. (This also applies to Pirani gauges, although the sensing filament is at a much lower temperature.) The external surfaces of the gauge heads can reach high temperatures and suitable shrouding should be provided if accidental contact is likely.

During degassing procedures there is intense light emission from glass gauge heads and dark goggles should be worn if it is necessary to view the gauge head. Screening should be arranged to protect passers-by. Gauge heads of the glass type should have connecting leads supported to prevent stress on the glass.

The electrodes of these gauges often have exposed multiconnection pins. Do ensure that common electrodes are properly electrically insulated.

Mechanical vacuum pumps

Electrical wiring
1. Ensure correct direction of rotation (three-phase motor).
2. Thermal overload cut-out switches for motors should be used.

Contact with moving parts
1. Guards are provided to prevent accidental contact with moving parts and pumps must never be operated with these guards removed.
2. Mechanical booster pumps should not be run in such a way that human contact with the rotating impellers is possible.

Oil mist hazard
Rotary pumps are oil sealed and discharge small quantities of oil mist when operating. In poorly ventilated areas this can lead to unacceptable concentrations of oil vapour. Oil vapour should be kept to a minimum. It is strongly recommended that a suitable oil mist filter is fitted to the pump or, alternatively, the exhaust should be piped away externally.

Exhaust hazard
Any dangerous substances which may be present in the system being evacuated may pass through the system into the pump and be emitted at the exhaust. Suitable precautions must be taken.

Dangerous substances
Hazardous situations will occur if pumps are used to pump flammable or corrosive substances, unless the pumps have been supplied specifically for the particular application.

Hazardous overpressure
1. Hazardous pressures may be produced if the exhaust from a mechanical pump is restricted or blocked, and care should be taken at all times to prevent this.
2. Exhaust line manifold systems should be designed to cope with the maximum exhaust load that can occur.
3. Hazardous pressures may be produced in the system to be evacuated if the rotary pump is operated in the reverse sense. When a rotary pump motor requires a three-phase electrical supply, the outfit should be checked for correct rotation after mains connection, before being connected to the system to be evacuated.

Incorrect mounting
Hazardous situations will occur due to the mass of the pump, if it is not supported by a structure of sufficient strength, or if the pump is not properly secured, using antivibration mountings where appropriate.

Suction applications
When mechanical pumps are used to provide suction for medical purposes, precautions must be taken to ensure that the suction and exhaust connections cannot be accidentally reversed.

Mechanical pump accessories

Traps, filters and condensers
If any hazardous material accumulates in a trap or filter then a hazard exists when the content of the trap is removed and during disposal of the trapped material.

Foreline traps
1. These traps use activated alumina pellets to trap oil vapour and thus prevent migration of pump oil to other parts of the vacuum system.
2. Activated alumina is a desiccant and inhalation of its dust, eye contact and unnecessary skin contact should be avoided.
3. If reactivation of the alumina is attempted by baking, then any other materials that have been absorbed will be released and suitable precautions will be necessary.

Inlet desiccant traps
1. These traps use phosphorus pentoxide (P_2O_5) as a desiccant.
2. It is highly caustic and reacts violently with water, evolving heat and forming a mild acid. Handle carefully to avoid creating a dust. Avoid inhaling dust or fumes.
3. Rubber gloves and eye protection should be worn when handling P_2O_5, especially when washing trays of partially spent P_2O_5.
4. P_2O_5 attacks aluminium; spillage on the body of the trap should therefore be avoided.
5. If any is spilt on the skin or clothing it should be washed off with plenty of water.
6. Full health and safety information on P_2O_5 should be obtained from the supplier.

Inlet chemical traps
1. These traps use activated charcoal to protect the pump from heavy lacquer and chemically active vapours and gases.
2. Precautions appropriate to the trapped materials should be taken when removing or disposing of the spent charcoal.

Oil mist filters
1. These filters are intended to remove oil mist from rotary pump exhaust. They should not be relied on to remove any other hazardous material.
2. They must not be allowed to become blocked, as hazardous overpressures could then occur.

Diffusion and vapour booster pumps

Hot surfaces and heat hazards
1. The external surfaces of vapour diffusion and booster pumps reach high temperatures. Electric cables in close proximity to these pumps must have insulating properties compatible with the temperatures reached.
2. These pumps should not stand on wooden floors or on surfaces of heat-sensitive or combustible material.
3. They should be installed in such a position that accidental bodily contact

with the hot surfaces is prevented, and adequate room should be allowed around the pump boiler for air circulation to take place.

Hot vapour hazards
1. If a hot vapour pump is opened to the atmosphere, vapour will be released, and could be a health hazard.
2. Fluid should not be drained from pumps or the pumps opened to atmosphere before they are cool. Heating oils at atmospheric pressures can cause them to decompose, burn or explode.

Cooling water failure
Lack of cooling water can cause pump overheating and steam generation in the cooling lines.

Mercury pump hazards
1. Mercury vapour pumps have the additional hazards associated with mercury. Mercury poisoning is cumulative and causes irreversible brain damage and fatal kidney damage. Do not expose mercury to the atmosphere at elevated temperatures ($>23\,°C$) since the vapour pressure increases rapidly with temperature with a corresponding increase in the hazards.
2. Mercury vapour pumps should only be fitted to vacuum systems designed to be resistant to mercury attack.
3. Any spillage must be immediately cleaned, by mechanical means, e.g. vacuum probe, followed by conversion of the remaining residue into mercury sulphide, so eliminating the danger of vaporization.
4. Mercury must be contained in tightly stoppered containers, preferably unbreakable.

Perfluoropolyethers (PFPE)
PFPE fluids produce toxic breakdown products when heated above 280 °C. Their use in diffusion pumps is not recommended.

Vapour traps and baffles

Condensate hazard
Condensible vapours present in a vacuum system will condense and accumulate on the cold surfaces of vapour traps. The material will then re-evaporate when the trap or baffle warms up. If the material is hazardous, suitable precautions should be taken.

Hazards of liquid nitrogen cooled traps and baffles
1. Only equipment designed for the purpose should be used for transferring liquid nitrogen from its storage container to the trap.

2. Skin contact with liquid nitrogen or surfaces at liquid nitrogen temperature will result in cold burns, and protective clothing should be worn where there is such a risk.
3. Spillage of liquid should be avoided by using suitable filling devices. Full-face protection should be worn where and when there is a risk of liquid splashing.
4. Filling funnels and open dewars should always be kept below shoulder level when filling traps.
5. Traps intended to be charged with liquid nitrogen should not be charged with other substances. In particular, they should never be charged with liquid air or liquid oxygen which could create a fire/explosion hazard.
6. The reservoirs of liquid nitrogen traps must always allow the escape of evaporating nitrogen gas; otherwise there will be a dangerous buildup of internal pressure and eventual explosion. The filling tube should only be covered with the ventilated cap provided and should never be blocked with, for example, a bung.
7. The filling spouts of cold traps should not be positioned where water can drip on them as this could cause the spouts to become blocked with ice.
8. Evaporated nitrogen can displace air and in confined spaces could possibly lead to asphyxia.
9. Cold surfaces at liquid nitrogen temperatures can condense certain permanent gases such as CO_2, xenon and argon. Dangerous pressures could be produced by a trap that has condensed quantities of such gases if it subsequently warms up on a sealed system. If such gases are passed through the trap it should be evacuated while it is warming up.
10. The cold surface of liquid nitrogen traps should not be exposed to the atmosphere while cold, as oxygen from the air can be condensed with the consequent fire/explosion hazards.

Turbomolecular pumps

The pumps are not normally designed to run at or near atmospheric pressure and certainly should not be run or rotated with the inlet exposed at atmospheric pressure without a shield on the inlet.

Cryopumps (see additionally Section 9.6)

1. Cryopumps can condense and accumulate all atmospheric gases and vapours.
2. If hazardous materials are present in the vacuum system and are retained by the cryopump, suitable precautions must be taken when the pump is warmed up.
3. A pressure relief valve is provided to relieve overpressure in the event of the cryopump warming up when the system is sealed. A dangerous

situation could occur if this valve failed to operate or if it operated when hazardous materials were present in the pump.
4. Cryopumps should not be operated when open to atmosphere.

Sputter-ion pumps

High-voltage and magnetic hazards
1. These devices operate from a high-voltage supply.
2. Connection and disconnection of the pump and power supply should only be undertaken with the power supply disconnected from the mains.
3. It is recommended that an additional protective earth conductor is connected to the pump body.
4. Powerful magnets form part of these pumps. Any equipment that is sensitive to a magnetic field should be kept away from them and if the magnets are removed care should be taken to avoid trapping fingers between the magnet and any ferromagnetic material or other magnets.

Fluids

Pump oils
1. Mineral-oil-based products are only slightly to moderately irritating to the skin and eyes. Prolonged exposure of the skin to mineral oils may give rise to dermatitis.
2. Care should, however, be taken to avoid inhalation of vapours or mists arising from undue heating or excessive mist generation. In the case of fluorinated compounds (e.g. Fomblin) avoid contact with excessive heat ($\geqslant 280\,°C$), e.g. lighted cigarettes, heater elements, etc.
3. Storage. Normally these products require no special fire precautions but it is recommended practice to store away from heat. When heat is required to facilitate handling of the product, this should be kept to a minimum.
4. Fluid spillage should be absorbed with sand, earth or mineral absorbent and disposed of in the proper manner. In the event of large spillages, steps should be taken to prevent pollution of drainage systems, rivers or waterways.
5. Where oxygen or aggressive materials have to be pumped, an inert and chemically resistant lubricant must be considered. Suitable members of fully fluorinated oils can be used.
6. The operating temperature of most vapour pump fluids is of the order of 195 to 220 °C. This constitutes a safe operating condition in a vacuum environment. However, a hazardous condition with some fluids may occur if the fluid is permitted to overheat excessively by allowing it to operate at or near atmospheric pressure. Under these conditions, the fluid temperature may approach the autoignition point (depending on the fluid, in the range from 300 to 600 °C).

Solvents
Trichloroethane is the preferred cleaning solvent; it has a low toxicity and is non-flammable. However, in common with all highly volatile oil and grease solvents, the main hazard is the narcotic, anaesthetic and toxic effects of breathing concentrations of vapour resulting from its use in confined spaces. Ensure there is adequate ventilation and rate of air change in the cleaning area. Smoking is prohibited. Avoid skin contact.

Pipeline valves

Pneumatic operation
1. Care should be taken to avoid trapping the fingers in pneumatically operated valve mechanisms.
2. Air pressures in excess of the operating range specified must not be applied to the inlet of pneumatic mechanisms.
3. The compressed air supply used should incorporate suitable pressure controllers and relief valves.

Electrical leadthroughs

General
1. These leadthroughs are normally provided with functional insulation only. Any necessary enclosure or insulation to protect against electric shock must be additionally provided.
2. Avoid hanging heavy wire leads from electrical leadthroughs as localized stress may cause a fracture.

Leak detection equipment

High-frequency spark testers
1. These units generate a high-frequency, high-voltage spark from an accessible unshielded probe.
2. They can ignite flammable material and should not therefore be used where flammable material is present.
3. Bodily contact with the spark should be avoided at all times and the spark testers should only be used by responsible trained personnel.
4. A serious electric shock is unlikely to be received if the probe is accidentally touched while operating, but injury could occur if the spark reached sensitive parts of the body (e.g. eyes). There could be a risk to persons suffering from heart weakness.
5. Injury could also occur as a result of the reflex movement after spark contact if the tester is used in the vicinity of other hazards, such as hot surfaces or rotating machinery.

6. Spark testers are used to test evacuated glass systems. Suitable eye protection must be worn during testing.

Vacuum systems and process plant

Hoists

Hydraulically operated hoists can become hazardous due to incorrect maintenance or damage. They should have periodic checks for loss of hydraulic fluid and for jerky operation, which is an indication of air leakage into the hydraulic system.

Exposure of human body to vacuum

1. No part of the human body should be exposed to vacuum.
2. Evacuation of the whole body will result in death and the exposure of small areas of the body surface to the suction of the pumping system can result in tissue damage.
3. Where large vacuum chambers have to be entered by personnel for loading, cleaning or maintenance reasons, procedures must be designed to prevent accidental evacuation of the workchamber with personnel inside.

Excess pressure hazard when backfilling

1. Unless otherwise stated vacuum systems are not intended to be raised to a pressure greater than that of the external atmosphere, and a hazardous situation will arise if a chamber is accidentally put under positive pressure.
2. Processes sometimes require a vacuum chamber to be filled with a gas other than atmospheric air. If the source of this gas is a pressure cylinder then pressure-reducing devices and safety valves should be incorporated in the line to avoid pressurizing the vacuum system.

Implosion hazards

1. Glass components are normally designed with a large safety factor. Failure is normally due to flaws in the material or through misuse. Examples of misuse include where the glass:
   (a) is knocked (by a handtool, etc.),
   (b) is not properly supported (e.g. when opening/closing glass valves),
   (c) has localized hotspots or cold areas due to the process,
   (d) is constrained by metal clamping.
2. Glass vacuum chambers are provided with implosion guards to contain any flying debris resulting from implosion. Such chambers should not be evacuated without the implosion guard in position. Wear safety glasses.
3. Glass vacuum chambers should always be checked for damage that might weaken them and cause implosion.

Cleaning hazards
1. Suitable precautions should be taken to protect personnel engaged in cleaning vacuum systems from the solvents being used and from process debris (dust) in the system. This particularly applies to personnel who enter large vacuum chambers to clean them.
2. The exact precautions to be taken will be dependent on the cleaning solvent involved but particular attention should be paid to the danger of inhaling solvent vapours (see 'Solvents', page 285).

Evaporation sources
1. Precautions should be taken to protect the eyes of both users and passers-by from the intense light emission of hot evaporation sources used in coating units.
2. When it is necessary to view a hot evaporation source, dark safety goggles should be worn.

Vacuum coating—deposits
The deposit on the chamber walls of a vacuum coating system are generally in the form of extremely fine particles. The nature, as well as the form of the materials, poses the following potential hazards:

1. Inhaling fine particles (powder) may cause damage to the lungs. Wear a protective respirator mask with a fine filter to help prevent this.
2. Some substances are toxic and inhaling them should be avoided. Take steps to ascertain whether or not the material being deposited is a known toxic substance.
3. Certain powders, titanium, for instance, and unoxidized aluminium particles, can cause flash fires when exposed to oxygen or other oxidizers. Therefore, when opening the chamber door after a deposition cycle, exercise extreme caution and allow time for the coating surface to oxidize. Breakage of some of the deposits of the more reactive condensates may still be hazardous, even with the above precautions. In this situation fire-protective clothing should be worn.

LT sources
1. Evaporation sources are supplied from low-voltage power supplies capable of delivering high currents in normal use.
2. Accidental short circuiting of these supplies is a serious hazard.
3. Cables carrying such currents must be protected from mechanical damage.

High-voltage supplies
1. Some vacuum processes involve the supply of high voltage to the vacuum system for HT discharge cleaning, electron gun use, etc.
2. The HT supplies should be interlocked with switches on the vacuum

system to prevent high voltage being available when the workchamber is open at atmospheric pressure.

Chamber air admittance hazards
1. Air admission to vacuum systems should always be carried out in a controlled manner, to avoid excessive disturbance of the surrounding atmosphere.
2. Large quarter-swing valves should not be used as air admittance valves unless a suitable flow restrictor is fitted. Too rapid air admission can cause violent pressure reductions in the room and violent air movement, with consequent risk of injury.
3. Large vacuum chambers should not be placed in small inadequately ventilated rooms where air admittance could cause a fall in room pressure.

Rotary workholders
To facilitate loading, rotary workholders can be operated when accessible, and care is necessary to avoid loose clothing or parts of the body being trapped by moving parts.

Appendix A

A selected bibliography

Vacuum measurement

Fitch, R. K. (1987) 'Total pressure gauges', *Vacuum*, **37**, 637–641.
Nash, P. (1987) 'The use of hot filament ionisation gaugess', *Vacuum*, **37**, 643–649.
Poulter, K. F. (1981) 'Developments and trends in vacuum metrology', *Le Vide*, **207**, 521–530.
Redhead, P. A. (1984) 'The measurement of vacuum pressures', *J. Vac. Sci. Technol.*, **A2**(2), 132–138.
Steckelmacher, W. (1987) 'The calibration of vacuum gauges', *Vacuum*, **37**, 651–657.

Identification of gases present

Batey, J. H. (1987) 'Quadrupole gas analysers', *Vacuum*, **37**, 659–668.
Drinkwine, M. J., and Lichtman, D. (1981) 'Partial pressure analysers and analysis', American Vacuum Society Monograph Series.
Edwards High Vacuum International (1987) 'Intelligent ion gauges', Leaflet, Publication 07-D393-20-895.
James, A. P. (1987) 'Automation of residual gas analysers', *Vacuum*, **37**, 677–680.
Mao Fu Ming, and Leck, J. H. (1987) 'The quadrupole mass spectrometer in practical operation', *Vacuum*, **37**, 669–675.

Rotary pumps

Balfour, D., *et al.* (1984) 'Pumping systems for corrosive and dirty duties', *Vacuum*, **34**, 771–774.
Connock, P., Devaney, A., and Currington, I. (1981) 'Vacuum pumping of aggressive and dust laden vapours', *J. Vac. Sci. Technol.*, **18**(3), 1033–1036.
Currington, I., Devaney, A., and Connock, P. (1982) 'Mechanical vacuum pumping equipment for applications involving corrosive and aggressive materials', *J. Vac. Sci. Technol.*, **20**(4), 1019–1022.
Hablanian, M. H. (1981) 'Ultimate pressure of mechanical pumps and the effectiveness of foreline traps', *J. Vac. Sci. Technol.*, **18**(3), 1156–1159.
Harris, N. S. (1978) 'Rotary pump back-migration', *Vacuum*, **28**, 261–268.
Kearney, K. M. (1988) 'Careless pump maintenance will cost you', *Semiconductor International*, September, 77–81.
Sarkozy, R. F., and Gatti, H. W. (1986) 'Semiconductor vacuum processing: protection of the mechanical pumping system', *Solid State Technol.*, September, 143–149.
Wycliffe, H., and Power, B. D. (1981) 'Pumped oil feed systems for rotary pumps', *J. Vac. Sci. Technol.*, **18**(3), 1160–1163.
Wycliffe, H. (1987) 'Rotary pumps and mechanical boosters—as used on today's high vacuum systems', *Vacuum*, **37**, 603–607.

Oil-free mechanical pumps

Balzers High Vacuum Limited. 'Roots Vacuum pumps', Leaflet, Publication PP800014PE.
Berges, H-P., and Götz, D. (1988) 'Oil-free vacuum pumps of compact design', *Vacuum*, **38**, 761–763.
Budgen, L. J. (1982) 'A mechanical booster for pumping radioactive and other dangerous gases', *Vacuum*, **32**, 672–679.
Dennis, N. T. M., Budgen, L. J., and Laurenson, L. (1981) 'Mechanical boosters on clean or corrosive applications', *J. Vac. Sci. Technol.*, **18**(3), 1030–1032.
Hablanian, M. B. (1988) 'The emerging technologies of oil-free vacuum pumps', *J. Vac. Sci. Technol.*, **A6**(3), 1177–1182.
Laurenson, L., and Turrell, D. (1988) 'The performance of a multistage dry pump operating under non-standard conditions', *Vacuum*, **38**, 665–668.
Singer, P. H. (1988) 'Matching vacuum pumps to processes', *Semiconductor International*, September, 71–75.
Wong, L., *et al.* (1988) 'An evaluation of the composition of the residual atmosphere above a commercial dry pump', *J. Vac. Sci. Technol.*, **A6**(3), 1183–1186.
Wycliffe, H. (1987). 'Mechanical high-vacuum pumps with an oil-free swept volume', *J. Vac. Sci. Technol.*, **A5**(4), 2608–2611.
Wycliffe, H., and Salmon, A. (1971) 'The application of hydrokinetic drives to high vacuum mechanical boosters (Roots pumps)', *Vacuum*, **21**, 223–229.

Diffusion pumps

Colwell, B. H. (1980) 'Vapour pumping groups', *Vacuum*, **30**, 321–327.
Hablanian, M. H. (1974) 'Diffusion pump technology. Proceedings of 6th International Vacuum Congress, *Japan. J. Appl. Phys.*, Suppl. 2, Pt. 1, 25–31.
Hablanian, M. H. (1974) 'Diffusion pumps and their operation', *Solid State Technol.*, December, 37–45.
Hablanian, M. H. (1983) 'Diffusion pumps—performance and operation', American Vacuum Society Monograph Series.
Harris, N. S. (1977) 'Diffusion pump back-streaming', *Vacuum*, **27**, 519–530.
Laurenson, L. (1987) 'Diffusion pumps and associated fluids', *Vacuum*, **37**, 609–614.

Turbomolecular pumps

Abbel, K., Henning, J., and Lotz, H. (1982) 'New turbomolecular pumps for application with radioactive gases, e.g. tritium', *Vacuum*, **32**, 623–625.
Balzers High Vacuum Limited. 'Description, design fundamentals, examples of applications and performance of turbomolecular pumps', Leaflet, Publication PM800049PE.
Fischer, K., Henning, J., Abbel, K., and Lotz, H. (1982) 'Pumping of corrosive or hazardous gases with turbomolecular and oil-filled rotary vane backing pumps', *Vacuum*, **32**, 619–621.
Goetz, D. G. (1982) 'Large turbomolecular pumps for fusion research and high-energy physics', *Vacuum*, **32**, 703–706.
Henning, J. (1978) 'Trends in the development and use of turbomolecular pumps', *Vacuum*, **28**, 391–398.
Henning, J. (1988) 'Thirty years of turbomolecular pumps: a review and recent developments', *J. Vac. Sci. Technol.*, **A6**(3), 1196–1202.

Henning, J. (1980) 'A comparison of the construction of old and new turbomolecular pumps', *Vacuum*, **30**, 183–187.
Hucknall, D. J., and Goetz, D. G. (1987) 'Turbomolecular pumps', *Vacuum*, **37**, 615–620.
Knecht, T. A. (1977) 'A new generation of turbo vacuum pumps', *Research/Development*, **28**(3), 57–66.

Cryopumps

Bently, P. D. (1980) 'The modern cryopump', *Vacuum*, **30**, 145–158.
Edwards High Vacuum International (1986) 'Cryopumping', Leaflet, Publication 05-B523-01-895.
Hands, B. A. (1987) 'Cryopumping', *Vacuum*, **37**, 621–627.
Scholl, R. A. (1983) 'Cryopumping in semiconductor applications', *Solid State Technol.*, December, 187–190.
Welch, K. M., and Flegal, C. (1978) 'Helium cryopumping', *Industrial Research/Development*, March, 83–88.

Ion pumps

Audi, M. (1988) 'Pumping speed of sputter ion pumps', *Vacuum*, **38**, 669–671.
Audi, M., and Simon, M. de (1987) 'Ion pumps', *Vacuum*, **37**, 629–636.

Fluids

Laurenson, L. (1977) 'Perfluoropolyethers as vacuum pump fluids', *Research/Development*, November, 61–72.
Laurenson, L. (1982) 'Technology and applications of pumping fluids', *J. Vac. Sci. Technol.*, **20**(4), 989–995.
Mastroianni, M. J., Tarplee, M. C., and Gilbert, L. (1985) 'Vacuum pump fluids for semiconductor processing', *Semiconductor International*, November, 62–64.
O'Hanlon, J. F. (1981) 'Mechanical pump fluids for plasma deposition and etching systems', *Solid State Technol.*, October, 86–89.
O'Hanlon, J. F. (1984) 'Vacuum pump fluids', *J. Vac. Sci. Technol.*, **A2**(2), 174–181.

Vacuum systems

Harris, N. S. (1981) 'Practical aspects of constructing, operating and maintaining rotary vane and diffusion-pumped systems', *Vacuum*, **31**, 173–182.
Heppell, T. A. (1987) 'High vacuum pumping systems—an overview', *Vacuum*, **37**, 593–601.
Hoffman, D. (1979) 'Operation and maintenance of a diffusion-pumped vacuum system', *J. Vac. Sci. Technol.*, **16**(1), 71–74.

Vacuum system components/materials/cleaning

Adam, H., and Jokisch, G. (1987) 'Vacuum valves and their use in practice', *Vacuum*, **37**, 681–689.

Halliday, B. S. (1987) 'An introduction to materials for use in vacuum', *Vacuum*, **37**, 583–585.
Halliday, B. S. (1987) 'Cleaning materials and components for vacuum use', *Vacuum*, **37**, 587–591.

Leak detection

Becker, W., Huber, W. K., Moll, E., and Rettinghaus, G. (1977) 'A novel leak detector with turbomolecular pump'. *Proceedings of 7th International Vacuum Congress and 3rd International Conference on Solid Surfaces*, pp. 203–206.
Edwards High Vacuum International (1988). Mass spectrometer leak detectors', Leaflet, Publication 17-D154-31-895.
Reich, G. (1987) 'Leak detection with tracer gases; sensitivity and relevant limiting factors', *Vacuum*, **37**, 691–698.

Safety

Bevis, L. C., Harwood, V. J., and Thomas, M. T. (1975) 'Vacuum hazards manual', American Vacuum Society Monograph Series.
Duval, P. (1988) 'Problems in pumping aggressive, poisonous and explosive gases', *Vacuum*, **38**, 651–658.
O'Hanlon, J. F., and Fraser, D. B. (1988) 'American Vacuum Society recommended practices for pumping hazardous gases', *J. Vac. Sci. Technol.*, **A6**(3), 1226–1254.

Books on vacuum technology

(Note that the dates given are when the book was first published. Many of these books are now out of print but may be available from libraries.)

American Vacuum Society (1984) *Dictionary of Terms for Areas of Science and Technology Served by the AVS*, 2nd edition.
Barrington, A. E. (1963) *High Vacuum Engineering*, Prentice-Hall.
Berman, A. (1985) *Total Pressure Measurements in Vacuum Technology*, Academic Press.
Carpenter, L. G. (1970) *Vacuum Technology*, Adam Hilger.
Chambers, A., Fitch, R. K., Halliday, B. S. (1989) *Basic Vacuum Technology*, Adam Hilger.
Dennis, N. T. M., and Heppell, T. A. (1968) *Vacuum System Design*, Chapman and Hall.
Dushman, S., and Lafferty, J. M. (1962) *Scientific Foundations of Vacuum Technique*, John Wiley.
Guthrie, A. (1963) *Vacuum Technology*, John Wiley.
Harris, N. S. (1977) *Vacuum Engineering*, AUEW-TASS Monograph.
Holland, L., Steckelmacher, W., and Yarwood, J. (1974) *Vacuum Manual*, E. and F. N. Spon.
Leck, J. H. (1957) *Pressure Measurement in Vacuum Systems*, Chapman and Hall.
Lewin, G. (1965) *Fundamentals of Vacuum Science and Technology*, McGraw-Hill.
Lewin, G. (1987) *An Elementary Introduction to Vacuum Technique*, AVS Monograph.
Leybold GmbH (1982) *Vacuum Technology—Its Foundations, Formulae and Tables*.

O'Hanlon, J. F. (1989) *A User's Guide to Vacuum Technology*, 2nd edition, John Wiley.

Pirani, M., and Yarwood, J. (1961) *Principles of Vacuum Engineering*, Chapman and Hall.

Power, B. D. (1966) *High Vacuum Pumping Equipment*, Chapman and Hall.

Redhead, P. A., Hobson, J. P., and Kornelson, E. V. (1968) *The Physical Basis of Ultra-high Vacuum*, Chapman and Hall.

Roberts, R. W., and Vanderslice, T. A. (1963) *Ultrahigh Vacuum and Its Applications*, Prentice-Hall.

Roth, A. (1976) *Vacuum Technology*, North-Holland.

Ryans, J. L., and Roper, D. L. (1986) *Process Vacuum System Design and Operation*, McGraw-Hill.

Spinks, W. S. (1963) *Vacuum Technology*, Chapman and Hall.

Steinherz, H. A. (1963) *Handbook of High Vacuum Engineering*, Reinhold Publishing.

Turnbull, A. H., Barton, R. S., and Riviere, J. C. (1962) *An Introduction to Vacuum Technique*, George Newnes.

Van Atta, C. M. (1965) *Vacuum Science and Engineering*, McGraw-Hill.

Varian Associates, Inc. (1980) *Introduction to Helium Mass Spectrometer Leak Detection*.

Wutz, M., Adam, H., and Walcher, W. (1989). *Theory and Practice of Vacuum Technology*, Vieweg & Sohn.

Ward, L., and Bunn, J. P. (1967) *Introduction to the Theory and Practice of High Vacuum Technology*, Butterworths.

Weston, G. F. (1985) *Ultrahigh Vacuum Practice*, Butterworths.

Various Associates, Inc. (1986) *Basic Vacuum Practice*.

General sources

Additional sources of information on aspects of high and ultra-high vacuum are the annual transactions of the American Vacuum Society published variously by Macmillan, Pergamon Press and the American Institute of Physics; the transactions of the triennial meetings of the International Vacuum Congress; and the journals *Vacuum* (English), *Le Vide* (French), *Vakuum Technik* (German), *Vuoto* (Italian) and *Journal of Vacuum Science and Technology* (American), *Vakuum in der Praxis* (German).

Appendix B

Glossary

The following extracts are reproduced by permission of the British Standards Institution (BSI) and are taken from standards documents ISO 3529: Parts 1, 2 and 3, (1981). Complete copies of these standards can be obtained from national standards bodies.

Back-migration
(a) In the case of the fluid entrainment pump, the passage of the pump fluid into the vessel to be evacuated, due to migration of pump fluid molecules on the surfaces.
(b) In the case of oil-sealed vacuum pumps; the passage of the pump oil within the vessel to be evacuated, due to migration of oil molecules on the surfaces.

Back-streaming of pump fluid   The passage of the pump fluid through the inlet port of the fluid entrainment pump (or of any associated baffle or trap) in a direction opposite to the direction of desired gas flow.

Baffle   A system of screens, possibly cooled, placed near the inlet of a vapour jet or diffusion pump, to reduce back-streaming and back-migration.

Cold trap   A trap that operates by condensation on cooled surfaces.

Conductance   Of a duct, or part of a duct, or an orifice: the throughput divided by the difference in mean pressures prevailing at two specified cross-sections or at both sides of the orifice, assuming isothermal conditions.

Critical backing pressure   The backing pressure above which a vapour jet or diffusion pump fails to operate correctly. It is the highest value of the backing pressure at which a small increment in the backing pressure does not yet produce a significant increase of the inlet pressure. The critical backing pressure of a given pump depends mainly on the throughput. (Note that for some pumps the failure does not occur abruptly and the critical backing pressure cannot then be precisely stated.)

Cryopump   An entrapment pump consisting of surfaces refrigerated to a low temperature sufficient to condense residual gases. The condensate is then maintained at a temperature such that the equilibrium vapour pressure is equal to or less than the desired low pressure in the chamber. (Note that the temperature chosen shall be in the range below 120 K depending on the nature of the gases to be pumped.)

Degassing   The deliberate desorption of gas from a material.

Diffusion pump   A kinetic pump in which a low-pressure, high-speed vapour stream provides the entrainment fluid. The gas molecules diffuse into this stream and are driven to the outlet. The number density of gas molecules is always low in the stream. A diffusion pump operates when molecular flow conditions obtain.

Gas ballast pump   A positive displacement pump in which a controlled quantity of a suitable non-condensable gas is admitted during the compression part of the cycle so as to reduce the extent of condensation within the pump.

Gauge head Of certain types of gauge, the part of the gauge that contains the pressure-sensitive element and is directly connected to the vacuum system.

Intermediate flow The passage of a gas through a duct under conditions intermediate between laminar viscous flow and molecular flow.

Magnetic deflection mass spectrometer A mass spectrometer in which accelerated ions are separated into different circular arcs under the action of a magnetic field.

Mass spectrometer An instrument that separates ionized particles of different mass-to-charge ratios and measures the respective ion currents. (Note that the mass spectrometer may be used as a vacuum gauge to measure the partial pressure of a specific gas, as a leak detector sensitive to a particular search gas, or as an analytical instrument to determine the percentage composition of a gas mixture. Types are distinguished by the methods of separating the ions.)

Molecular flow The passage of a gas through a duct under conditions such that the mean free path is very large in comparison with the largest internal dimension of a transverse section of the duct.

Outgassing The spontaneous desorption of gas from a material.

Quadrupole mass spectrometer A mass spectrometer in which ions are injected axially into a quadrupole lens consisting of a system of four electrodes, usually rods, to which radio frequency and d.c. electric fields in a critical ratio are applied, so that only ions with a certain mass-to-charge ratio emerge.

Saturation vapour pressure The pressure exerted by a vapour which is in thermodynamic equilibrium with one of its condensed phases at the prevailing temperature.

Sensitivity For a given pressure, the change in the reading indicated by the vacuum gauge, divided by the corresponding change in pressure. (Note that for certain types of gauge, the sensitivity depends on the nature of the gas. In such a case, the sensitivity for nitrogen must always be specified.)

Sliding vane rotary vacuum pump A rotary positive displacement pump in which an eccentrically placed rotor is turning tangentially to the fixed surface of the stator. Two or more vanes, sliding in slots of the rotor (usually radial) and rubbing on the internal wall of the stator, divide the stator chamber into several parts of varying volume.

Throughput The quantity of gas (in pressure–volume units) passing through a cross-section in a given interval of time, divided by that time. It is also the mass flow rate divided by the unitary mass density.

Total pressure vacuum gauge A vacuum gauge for measuring the total pressure of a gas or a gaseous mixture. (Note that the compression gauge measures only the pressure of the gases that do not condense within the gauge during the measuring procedure.)

Turbomolecular pump A molecular drag pump in which the rotor is fitted with discs provided with slots or blades rotating between corresponding discs in the stator. The linear velocity of a peripheral point of the rotor is of the same order of magnitude as the velocity of the gas molecules. A turbomolecular pump operates normally when molecular flow conditions obtain.

Ultimate pressure of a pump The value towards which the pressure in a standardized test dome tends asymptotically, without introduction of gas and with the pump operating normally. A distinction may be made between the ultimate pressure due only to non-condensable gases and the total ultimate pressure due to gases and vapours.

Vacuum   A commonly used term to describe the state of a rarefied gas or the environment corresponding to such a state, associated with a pressure or a mass density below the prevailing atmospheric level.

Vacuum gauge   An instrument for measuring gas or vapour pressure less than the prevailing atmospheric pressure. (Note that some types of vacuum gauges commonly in use do not actually measure a pressure, as expressed in terms of a force acting on a surface, but some other physical quantity related to pressure, under specific conditions.)

Viscous flow   The passage of a gas through a duct under conditions such that the mean free path is very small in comparison with the smallest internal dimension of a traverse section of the duct, the flow being therefore dependent on the viscosity of the gas. The flow may be laminar or turbulent.

Volume flow rate of a vacuum pump   It is the volume flow rate of the gas removed by the pump from the gas phase within the evacuated chamber. This kind of definition is only applicable to pumps that are distinct devices, separated from the vacuum chamber. For practical purposes, however, the volume flow rate of a given pump for a given gas is, by convention, taken to be the throughput of that gas flowing from a standardized test dome connected to the pump, divided by the equilibrium pressure measured at a specified position in the test dome, and under specified conditions of operation.

Appendix C

Numbers written as powers of ten

A pressure of, say, 0.000 007 mbar might be recorded in a vacuum system. The figure 0.000 007 when written thus in the decimal system is clumsy to use, is subject to transcription errors and is likely to be misinterpreted when spoken. In vacuum use, such numbers are expressed conveniently as multiples of ten, e.g. 0.000 007 is written as 7×10^{-6} and spoken as 'seven times ten to the minus six'. Ten to the minus six, or more correctly ten to the power of minus six, is 0.000 001 or one-millionth.

As a further example, the number 0.005 06 is written as 5.06×10^{-3}. The conversion is achieved by moving the decimal point to the right, figure by figure, until a number between 1 and 10 appears, i.e. 0.005 06 becomes 0005.06. The number of figures by which the decimal point was moved is then counted—in this case three—and this number is written after the minus sign, i.e. 10^{-3}.

To convert, say, 6.3×10^{-4} back to a decimal, the number after the minus sign (the index) is used to determine how many places the decimal point must be moved to the left. In this case, four, giving 0.000 63. To add and subtract powers of ten, the numbers can be converted to decimals, the arithmetic done and the result converted back, e.g.

$$5.4 \times 10^{-6} + 7.25 \times 10^{-4}$$

$$= 0.000 005 4 +$$

$$0.000 725 0$$

$$\overline{0.000 730 4} = 7.304 \times 10^{-4}$$

or the numbers can be converted to the same power of ten, e.g.

$$5.4 \times 10^{-6} + 7.25 \times 10^{-4}$$

$$= 5.4 \times 10^{-6} + 725 \times 10^{-6}$$

$$= 730.4 \times 10^{-6}$$

$$= 7.304 \times 10^{-4}$$

Numbers greater than 1 may also be written as powers of ten, e.g.

$$10 000 = 10^{4}$$

(spoken 'ten to the power four')

$$1013 = 1.013 \times 10^{3}$$

$$25 \text{ million million} = 2.5 \times 10^{13}, \quad \text{etc.}$$

Table C.1 illustrates some examples of numbers written as positive and negative powers of ten.

Table C.1   Positive and negative powers of ten

Number		Power of ten
1000	$= 1 \times 10 \times 10 \times 10$	$= 1 \times 10^3$
100	$= 1 \times 10 \times 10$	$= 1 \times 10^2$
10	$= 1 \times 10$	$= 1 \times 10^1$
1	$= 1 \times 1$	$= 1 \times 10^0$
0.1	$= 1 \times \dfrac{1}{10}$	$= 1 \times 10^{-1}$
0.01	$= 1 \times \dfrac{1}{10 \times 10}$	$= 1 \times 10^{-2}$
0.001	$= 1 \times \dfrac{1}{10 \times 10 \times 10}$	$= 1 \times 10^{-3}$

Appendix D

Reading gauge scales in powers of ten

To read a gauge with a scale in powers of 10:

1. Look at the pointer and note marked scale numbers on either side of the pointer.
2. Decide which of these represents the smaller pressure.
3. Work out what pressures are represented within the unmarked divisions. Note that these will have same index as the lower pressure and their value will progressively increase from this lower value.
4. Work out the pressure indicated.

Figure D.1 shows this exercise carried out for a meter with typical scale markings:

Reading A = approximatey 7×10^{-3} mbar
Reading B = approximately 2.5×10^{-1} mbar
Reading C = between 5 mbar and atmosphere

Note that some instruments may have scales with pressure increasing from left to right as shown; others may read from right to left.

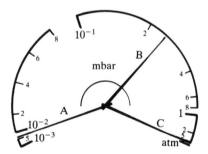

Figure D.1   Three examples (A, B and C) of gauge readings

Appendix E
Logarithms and logarithmic scales

In vacuum practice logarithms are encountered in graphs, showing, for example, how the pumping speed of a particular pump varies with pressure over a very wide range (see, for example, Figure 5.7).

If, say, a range of pressure was plotted from 1000 mbar down to 1 mbar on a linear scale with 1 mm representing 1 mbar, then 1 metre would be required to represent the whole range! This is clearly impractical and for this reason, instead of plotting a linear scale, it is normal to use a logarithmic scale, i.e. a scale based on the logarithms of the numbers involved.

A number such as 1000 can be written as 10^3; instead of writing the number in this way it is possible to express it by the logarithm 3 as long as it is understood that the number concerned is given by 10 raised to the third power. It is said that 3 is the logarithm of 1000 and this makes it possible to use small numbers (e.g. 3 or -3) to represent very large or very small numbers (e.g. 1000 or 0.001 respectively).

We can now construct a scale based on the logarithms of the numbers involved. Equally spaced divisions on the graph paper now represent units of logarithms; i.e. if one division were to represent the distance between the logarithm 1 and the logarithm 2 the next higher division would represent the logarithm 3. Consequently, whole logarithms form a linear scale. Each division on this scale represents a change of a factor 10 in the actual numbers represented.

Figure E.1 compares linear and logarithmic scales. In this figure a unit on the logarithmic scale has been taken arbitrarily to equal a distance equal to 9 mbar (1 to 10 mbar) on the linear scale. The zero point on the logarithmic scale represent 1 mbar. Actually this zero point is $\log_{10} 1 = \log_{10} 10^0 = $ logarithm 0.

The next division on the logarithmic scale would be $\log_{10} 10 = \log_{10} 10^1$ $= $ logarithm 1. Therefore, two successive logarithms have been arbitrarily chosen, viz. 0 and 1, to correspond on the linear scale to 1 to 10 mbar. The next division on the logarithmic scale is 2 or $\log_{10} 100$, and so on. Therefore, on the logarithmic scale shown it is possible to represent a change of pressure from 1 to 100 mbar, whereas the same distance on the linear scale only represents a pressure change of 1 to 28 mbar.

The compression in the logarithmic scale is clearly evident, e.g. the distance from 2 to 3, etc. Note that a whole logarithmic unit represents a change of factor of 10 (decade) in the number being plotted. The actual numbers are shown on the scale rather than the logarithms.

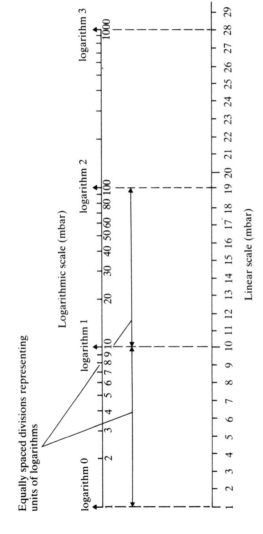

Figure E.1 Comparison of logarithmic and linear scales

Appendix F

Temperature scales

Fahrenheit, Celsius and Kelvin scales are all used to express temperature. The Kelvin or absolute scale is usually used for cryopump temperatures.

	K	°C	°F
Boiling point of water	373	100	212
Freezing point of water	273	0	32
Boiling point of liquid nitrogen	77	−196	−321
Lowest cold head temperature	10	−263	−441
Absolute zero of temperature	0	−273	−460

Note: $K = °C + 273$
$°C = 5/9 \ (°F - 32)$
$°F = 9/5 °C + 32$

Appendix G

Simple ideas of atomic structure

Definitions

1. *Element.* An element is a substance that cannot be split up into anything simpler by a chemical change.
   Elements may be solid like copper and carbon; liquid like mercury; gaseous like nitrogen and oxygen.
2. *Atom.* An atom is the smallest possible particle of an element that can exist.
3. *Molecule.* A molecule is the smallest part of a substance that can normally exist in the free state. It can consist of one or more atoms.

Atoms can be considered to be composed of three parts called *atomic particles*: electrons, protons and neutrons. The electrons are negatively charged particles moving in orbital paths around the nucleus of the atom. The nucleus contains two particles: a positively charged proton, its charge being equal but opposite to the charge on an electron, and an electrically neutral neutron.

Everything in the physical universe is composed of ninety-six stable elements and twelve unstable ones. Each element is composed of different combinations of the three atomic particles.

Fortunately, nature combined the atomic particles in an orderly manner:

1. There are equal numbers of protons and electrons in every atom.
2. Electrons move in specific orbits which can contain a maximum number of electrons.

The whole bulk of the atom, as defined by the outermost electron shell, is very great compared with the size of the nucleus (in analogy, about the same proportion as the dome of St Paul's to a clenched fist).

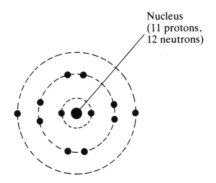

Nucleus
(11 protons,
12 neutrons)

Electrons revolving in orbits

Figure G.1 Magnesium atom (diagrammatic)

A numerical value is assigned to each element equal to the number of protons (and therefore the number of electrons) in the atom. It is called the *atomic number*. We know immediately that the element with atomic number 20 (calcium) has 20 protons and 20 electrons.

A diagrammatic representation of a typical atom is shown in Figure G.1. Note that the first orbit is filled at two electrons, whereas the second can hold eight. In general:

1. Elements with the same number of outer ring electrons have similar properties.
2. Elements tend to be stable with eight electrons in the outer ring (the first orbit is an exception).
3. Atoms tend to combine with other atoms to create eight electrons in the outer ring.

The simplest atom is that of hydrogen. It consists of a single proton with one electron rotating round it. In order of complexity, the simpler atoms are made up as shown in Table G.1, neutrons being omitted.

Table G.1 The electron shell structure of the simpler elements

Element	Atomic number = number of protons	Electrons in each shell		
Hydrogen	1	1		
Helium	2	2		
Lithium	3	2	1	
Beryllium	4	2	2	
Boron	5	2	3	
Carbon	6	2	4	
Nitrogen	7	2	5	
Oxygen	8	2	6	
Fluorine	9	2	7	
Neon	10	2	8	
Sodium	11	2	8	1
Magnesium	12	2	8	2
Aluminium	13	2	8	3
Silicon	14	2	8	4
Phosphorus	15	2	8	5
Sulphur	16	2	8	6
Chlorine	17	2	8	7
Argon	18	2	8	8
Potassium	19	2	8	8	1
Calcium	20	2	8	8	2

Another important number is the *mass number*. This is the total number of nucleons (=protons + neutrons); e.g. the mass number of lithuim is 7 (=3 protons + 4 neutrons).

Isotopes

The oxygen that we breathe has an atomic number 8; i.e. it has 8 protons in the nucleus (and 8 electrons to make a neutral atom). Most of this oxygen also has 8 neutrons (so the mass number = 8 + 8 = 16). However, some of the oxygen has 9 neutrons (but still 8 protons, making a mass number of 17). They have the same chemical properties but have different masses. They are called isotopes of oxygen. All elements have more than one isotope which cannot be separated by chemical means.

In order to separate them physical techniques such as mass spectrometry must be used (see Chapter 4).

Atomic masses

Atomic masses are measured on a relative scale on which the normal carbon atom has a mass of 12 atomic mass units (a.m.u.). Some examples are given in Tables G.2 and G.3.

Table G.2 Atomic masses of some elements $(1 \text{ a.m.u.} = 1.66 \times 10^{-27} \text{ kg})$

Element	Symbol	Mass (a.m.u.)
Hydrogen	H	1
Helium	He	4
Lithium	Li	7
Beryllium	Be	9
Boron	B	11
Carbon	C	12
Nitrogen	N	14
Oxygen	O	16
Fluorine	F	19
Neon	Ne	20
Aluminium	Al	27
Sulphur	S	32
Chlorine	Cl	35
Argon	Ar	40

Table G.3 Molecular formula and atomic masses of some substances commonly found in vacuum systems

Substance	Formula	Mass (a.m.u.)
Hydrogen	H_2	2
Ammonia	NH_3	17
Water	H_2O	18
Nitrogen	N_2	28
Carbon monoxide	CO	28
Nitric oxide	NO	30
Oxygen	O_2	32
Nitrous oxide	N_2O	44
Carbon dioxide	CO_2	44

Appendix H
Summary of formulae

For conductances, pumping speed, outgassing, ultimate pressure and pump-down time calculations used in Chapter 13

Conductance:

$$C = \frac{Q}{P_1 - P_2}$$

Parallel conductances:

$$C_{total} = C_1 + C_2 + C_3 + \cdots$$

Series conductances:

$$\frac{1}{C_{total}} = \frac{1}{C_1} + \frac{1}{C_2} + \frac{1}{C_3} + \cdots$$

Viscous conductance, circular tube:

$$C = \frac{136.5 D^4 P}{L}$$

Molecular conductance, circular tube:

$$C = \frac{12.1 D^3}{L}$$

Viscous conductance, orifice:

$$C = 20A$$

Molecular conductance, orifice:

$$C = 11.6A$$

Pumping speed (volume flow rate):

$$S = \frac{Q}{P}$$

Effective speed of a pump due to a conductance in series:

$$S_e = \frac{S_p \times C}{S_p + C}$$

Effective speed of a pump due to several conductances in series:

$$\frac{1}{S_{eff}} = \frac{1}{S_{pump}} + \frac{1}{C_1} + \frac{1}{C_2} + \dots \text{etc.}$$

Outgassing:

$$Q_{tot} = q_1 A_1 + q_2 A_2 + \ldots$$

Ultimate pressure:

$$P_{ult} = \frac{Q_{total}}{S_{eff}}$$

Pump-down time:

$$T = 2.3 \left(\frac{V}{s}\right) \log_{10}\left(\frac{P_1}{P_2}\right)$$

Index